Jianghua Zhang

Influence of Emitter Surface Roughness and Emission Inhomogeneity on Efficiency and Stability of High Power Fusion Gyrotrons

Karlsruher Forschungsberichte aus dem
Institut für Hochleistungsimpuls- und Mikrowellentechnik

Herausgeber: Prof. Dr.-Ing. John Jelonnek

Band 9

Influence of Emitter Surface Roughness and Emission Inhomogeneity on Efficiency and Stability of High Power Fusion Gyrotrons

by
Jianghua Zhang

KIT Scientific Publishing

Dissertation, Karlsruher Institut für Technologie (KIT)
Fakultät für Elektrotechnik und Informationstechnik, 2016
Referent: Prof. Dr.-Ing. John Jelonnek
Korreferent: Prof. Dr. Alan D. R. Phelps

Impressum

KIT Scientific Publishing

Karlsruher Institut für Technologie (KIT)
KIT Scientific Publishing
Straße am Forum 2
D-76131 Karlsruhe

KIT Scientific Publishing is a registered trademark of Karlsruhe Institute of Technology. Reprint using the book cover is not allowed.

www.ksp.kit.edu

Print on Demand 2016

ISSN 2192-2764
ISBN 978-3-7315-0578-5
DOI: 10.5445/KSP/1000058566

Influence of Emitter Surface Roughness and Emission Inhomogeneity on Efficiency and Stability of High Power Fusion Gyrotrons

Zur Erlangung des akademischen Grades eines

DOKTOR-INGENIEURS

von der Fakultät für

Elektrotechnik und Informationstechnik

des Karlsruher Institut für Technologie (KIT)

genehmigte

DISSERTATION

von

M.Sc. Jianghua Zhang

geb. in Henan, China

Tag der mündlichen Prüfung:	22. 07. 2016
Referent:	Prof. Dr.-Ing. John Jelonnek
Korreferent:	Prof. Dr. Alan D. R. Phelps

Foreword of the Editor

Gyrotrons are high-power millimeter-wave vacuum electron devices capable of generating Continuous-Wave (CW) power in the megawatt range. Major applications are the Electron Cyclotron Resonance Heating (ECRH) and Current Drive (ECCD) for heating and plasma stabilization of controlled thermonuclear fusion. For example, the ECRH system of ITER requires 1 MW of output power at 170 GHz in CW operation with 50 % total efficiency for a single gyrotron.

The electron beam quality is the key for any stable and efficient gyrotron operation. An insufficient electron beam quality causes a degraded gyrotron efficiency and output power. Additionally, it could cause instable operation of the gyrotron.

In his work, Dr.-Ing. Jianghua Zhang presents the influence of emitter surface roughness and emission inhomogeneity on the electron beam quality and operation of high power fusion gyrotrons. Those two factors are the main cause for the electron beam quality deterioration. An emitter surface roughness model, a new evaluation method for the emission inhomogeneity and an enhanced secondary electron emission model are developed. With the enhanced beam optics code which has been developed in frame of this work, the generation mechanisms for Low Frequency Oscillations (LFOs) and Electron Beam Halo (EBHs) are studied. The simulated generations of LFOs and EBHs in different gyrotron configurations are compared with experiments. In particular, the generation of EBH is verified for the first time. With the studied instability mechanisms using enhanced codes, three different methods to suppress the instabilities are proposed in this work.

Dr.-Ing. Jianghua Zhang is providing the gyrotron community new and very valuable tools to assist the advanced design of robust magnetron injection guns for gyrotrons in the future. It will help to develop high-power fusion gyrotrons with significantly higher efficiency and stability.

Kurzfassung

Der steigende Bedarf an leistungsfähigen, zuverlässigen und effizienten Leistungserzeugern (Gyrotrons) als Millimeterwellenquellen für die Elektronzyklotronresonanzheizung (ECRH) in Fusionsexperimenten erfordert einen tieferen Einblick in die verschiedenen Faktoren, welche die Leistungsfähigkeit von Gyrotrons bestimmen. In diesem Zusammenhang werden der Einfluss der Oberflächenrauhigkeit, der Inhomogenität, der Emission und der Erzeugung von Sekundärelektronen betrachtet. Speziell wird auf die unerwünschten Effekte der sogenannten „Low Frequency Oscillations“ (LFO) und des „Electron Beam Halo“ (EBH) eingegangen.

Ein neues Modell für die Emission von Elektronen bei einer bestimmten Oberflächenrauhigkeit des Emitters wird präsentiert, das in die KIT-Codes für Strahlsysteme (MIGs) ESRAY und ESPIC implementiert worden ist. Unter Verwendung dieses neuen Modells ist es möglich, den Einfluss der Oberflächenrauhigkeit von Emittern auf den Elektronenstrahlqualität und den Wirkungsgrad von Gyrotrons zu bestimmen. Um den Einfluss der zum Emitter zurück reflektierten Elektronen zu untersuchen, wird ein Modell für Sekundärelektronen gezeigt, das über eine hohe Genauigkeit bezüglich der Winkelauflösung für die hochenergetischen Sekundärelektronen (elastische und rückgestreute Elektronen) verfügt. Das Modell wurde in die KIT-eigenen Codes ESRAY und ESPIC implementiert. Zusätzlich wird eine neue Methode zu Bestimmung des Einflusses der Emissionshinhomogenität von Emittern entwickelt.

Unter Zuhilfenahme dieser neuen Werkzeuge wurde der Einfluss der Oberflächenrauhigkeit und der Emitterinhomogenität auf die Elektronenstrahlqualität und den Gyrotronwirkungsgrad numerisch untersucht. Die Simulationen zeigen, dass beide Faktoren die Elektronenstrahlqualität und den Wirkungsgrad von Gyrotrons signifikant beeinflussen. Es wurde herausgefunden, dass durch beide Faktoren reflektierte Elektronen entstehen können.

Unter Verwendung der neuen Möglichkeiten mit dem Programm ESPIC wurden zwei Phänomene untersucht: Raumladungs-LFOs und EBHs. Numerische Rechnungen, qualitative Analysen und experimentelle Ergebnisse des Kurzpulsprototyps für die EU 2 MW, 170 GHz koaxiale Röhre zeigen, dass Raumladungs-LFOs mit den TEM-Moden in der koaxialen Struktur des Gyrotrons verkoppeln können. Raumladungs-LFOs können somit den Betrieb eines Gyrotrons negativ beeinflussen. Die Erzeugung von EBHs in Gyrotron-MIGs wird in der Arbeit gezeigt. Die Korrelation zwischen EBHs und experimentell sichtbaren Spuren von Elektroneneinschlägen auf der Oberfläche von Gyrotronkomponenten wurde nachgewiesen. Basierend auf diesen untersuchten Mechanismen werden neue Designkriterien für Gyrotrons abgeleitet.

Das deutlich verbesserte Programm zur Berechnung der Strahloptik zusammen mit den untersuchten Phänomenen sind essentielle Bestandteile, um zukünftig bessere Designs für Gyrotron-MIGs zu erhalten.

Abstract

The increasing demand for powerful, reliable, and efficient gyrotron oscillators as millimetre-wave sources for Electron Cyclotron Resonance Heating (ECRH) in fusion plasma experiments requires a close look at the various factors in gyrotrons that determine gyrotron performance. In this frame, the influence of emitter surface roughness, emission inhomogeneity, and secondary electron generation on gyrotron operation is presented, with focus on Low Frequency Oscillations (LFOs) and Electron Beam Halo (EBH) generation, both being unwanted.

A new surface roughness model is developed and implemented into the KIT Magnetron Injection Gun (MIG) codes ESRAY and ESPIC. Using this new model, it is possible to determine the influence of surface roughness on the gyrotron's electron beam quality and efficiency. To investigate the influence of electrons reflected back to the emitter, a modified secondary electron emission model with high accuracy on the angular distribution of high energy secondary electrons (elastic and rediffused electrons) is developed and implemented in the codes ESRAY and ESPIC. In addition, a new evaluation method for the emission inhomogeneity is developed.

With these new tools, the influence of emitter surface roughness and of emission inhomogeneity on electron beam quality and gyrotron efficiency is investigated numerically. Simulations show that both factors affect the electron beam quality and gyrotron efficiency significantly. It was found that back-reflected electrons can be generated due to these two factors.

Two phenomena caused by back-reflected and secondary electrons in the gyrotron have been found: space charge LFOs and EBHs. Numerical calculation, qualitative analysis, and experimental results on the short-pulse prototype of the EU 2MW, 170 GHz coaxial-cavity gyrotron show that space charge LFOs can couple to a TEM mode in the coaxial-cavity gyrotron, and thereby can negatively affect stable operation of the gyrotron. The generation mechanism of EBHs in gyrotron MIGs is demonstrated. The correlation between EBHs and experimentally observed traces of electron impacts on the surface of gyrotron components is verified. Based on the mechanisms studied new design criteria for gyrotron MIGs are proposed.

The significantly enhanced beam optics code and analysed phenomena are important to assist in the advanced design of more robust MIGs for gyrotrons in the future.

Contents

List of used symbols, variables, constants and abbreviations

Symbols and variables

B	magnetic field
$\vec{B}$	magnetic field vector
B_m	maximum achievable magnetic field
d	distance
E	electric field
$\vec{E}$	electric field vector
E_{kin}	electron energy
$\vec{e}$	unit vector
f	frequency of an electromagnetic field or wave
f_{ce}	electron cyclotron frequency
f_{ci}	ion cyclotron frequency
I_b	beam current
$I_{\text{C-L}}$	space charge limited current
I_{R}	temperature limited current
J	current density
$J_{\text{C-L}}$	space charge limited current density
J_{R}	saturation current density
J_{S}	temperature limited current density
$K_{E_{\text{kin}}}$	energy coefficient

$K_{v_\perp}$	transverse velocity coefficient
K_r	surface roughness coefficient
k	wave number
$k_\perp, k_{\|\|}$	transverse and axial components of wave number
l	length
m_i	rest mass of the ion
m, n	mode index: azimuthal, radial
N_{ini}	number of electrons emitted from the emitter
N_{sec}	number of secondary electrons
N_{tot}	number of electrons
P_n	probability to generate n secondary electron
$P_{n,ts}$	probability to generate n true secondary electron
p	transverse momentum of electron
Q_{ini}	total space charge of electrons emitted from the emitter
Q_{sec}	total space charge of secondary electrons
Q_{tot}	total space charge
q_i	charge of ith electron
$q_{\text{ini},i}$	charge of ith initial electron
$q_{\text{sec},i}$	charge of ith secondary electron
r, θ, φ	cylindrical coordinates
r_0	surface microstructure size
r_L	Larmor radius
r_b	beam radius
r_{gc}	guiding centre radius
r_w	wall radius
S, dS	surface, surface element
s	harmonic factor of Ω_c

T	temperature
t	time
V_{acc}	accelerating voltage
V_b	beam voltage
V_B	body/anode voltage
V_C	cathode voltage
v	velocity
$v_\perp, v_\parallel$	transverse and axial components of velocity
W	work function
W_t	transition work function
x, y, z	Cartesian coordinates
Z_i	charge number of ion

Greek variables

α	pitch factor, i.e., the velocity ratio $\alpha = \beta_\perp/\beta_\parallel$
α_T	voltage factor in Tonnerre model
β	normalized velocity $\beta = v/c_0$
$\beta_\perp, \beta_\parallel$	transverse and axial components of β
$\beta_{\mathrm{ini},\perp}$	initial perpendicular velocity
$\beta_{\mathrm{C\text{-}L}}$	space charge limited current density correction factor
β_E	electric field enhancement factor
γ	relativistic factor (Lorentz factor)
δ	secondary electron emission yield
δ_E	electric field spread
δ_e	secondary electron emission yield of elastic secondary electrons
δ_J	emission inhomogeneity
δ_r	secondary electron emission yield of rediffused secondary electrons
δ_{ts}	secondary electron emission yield of true secondary electrons

$\sigma_{\hat{j}_{\mathrm{R}}}$	normalized effective emission inhomogeneity
η_{col}	collector efficiency
η_{elec}	interaction efficiency
η_{tot}	total efficiency
θ_i	fractional area
λ	wave length
μ	magnetic momentum
σ_T	standard deviation of temperature
σ_W	standard deviation of work function
φ	electric potential
Ω_c	electron-cyclotron angular frequency
ω	angular frequency of an electromagnetic field/wave, $\omega = 2\pi f$
χ_{mn}	n-th root of the Bessel function of order m

Constants

A_0, A_g	Richardson constant	1.20173×10^6 A/(m^{2}K^2), $0.5A_0$
$A_{\text{C-L}}$	Child-longmuir constant	2.334×10^{-6} A/V$^{3/2}$
c_0	speed of light	2.99792458×10^8 m/s
e	charge of an electron	$1.60217653 \times 10^{-19}$ C
k_B	Boltzmann constant	$1.3806505 \times 10^{-23}$ J/K
m_e	rest mass of the electron	$9.1093826 \times 10^{-31}$ kg
ε_0	vacuum permittivity	$8.854187817 \times 10^{-12}$ F/m

Abbreviations

CVC	current voltage characteristic
DEMO	DEMOnstration power plant
EBH	electron beam halo
EBOS	electron beam optics system

ECM	electron cyclotron maser
ECRH	electron cyclotron resonance heating
FDM	finite difference method
FEM	finite element method
ICRH	ion cyclotron resonance heating
IEA	International Energy Agency
IHM	Institute for Pulsed Power and Microwave Technology
LFO	low frequency oscillation
KIT	Karlsruhe Institute of Technology
MIG	magnetron injection gun
NBI	neutral beam injection
PIC	particle in cell
RF	radio-frequency
SDC	single-stage depressed collector
SEY	secondary emission yield
TE	transverse electric
TM	transverse magnetic
VED	vacuum electron device

1 Introduction

In this chapter, the overview of high power gyrotrons for thermal nuclear fusion is given first. Then, the research contents of this work are given after a review of challenges for fusion gyrotrons. At the end of this chapter, an introduction to the setup and work principles of gyrotrons are presented.

1.1 Introduction to thermonuclear fusion and fusion related heating systems

With the rapid growth of population and energy consumption per person but a limited amount of fossil fuels on the earth, people are looking for the clean energy for the next generations. Renewable energy and fusion are two potential solutions. However, renewable energy sources, such as solar power and wind, are expensive per unit of output; therefore, their consumption as a percentage of the total consumption was only 3.5 % in 2012 according to the report of International Energy Agency (IEA). Nuclear fusion, due to the practically limitless fuel supply and special energy generating mechanism could be an environmentally friendly power supply for the future.

Among the available techniques, magnetic confinement thermonuclear fusion is the most promising method to generate controllable power. In nuclear fusion, two light nuclei (typically the hydrogen isotopes ${}_{1}^{2}\mathrm{D}$ and ${}_{1}^{3}\mathrm{T}$) merge into one heavier nucleus (${}_{2}^{4}\mathrm{He}$), with the additional generation of neutrons [Law57]. The difference of rest mass before and after the fusion reaction is converted into energy. The generated amount of energy in fusion is based on the equation of mass and energy $E = \triangle mc^2$ which was proposed by Einstein [Ein00]. However, to merge two nuclei together, they should have sufficient kinetic energy to overcome the ion-ion electrostatic repulsive potential, the so called Coulomb barrier. In a magnetic confinement thermonuclear reactor, particles are heated up to 100-200 million °C; therefore, they can have sufficient kinetic energy to overcome the Coulomb barrier and reach a distance near enough to each other. At a sufficiently small distance, the strong attractive nuclear force dominates and the two nuclei fuse together [Har02].

There are generally three methods to heat up the ions in fusion plasmas [Har02]: ohmic heating, Neutral Beam Injection (NBI) [Kun80, HDG+09, IDPH+01], and Radio-Frequency (RF) plasma heating [Kam75]. The ohmic heating comes from the plasma current, which is induced in order to provide the poloidal magnetic field. This field is essential in plasma confinement in a tokamak. The ohmic heating works like an electric water heater where the resistor heats up when the current flow through it. However, this method is only used for the initial heating because the resistance decreases with increasing plasma temperature, which is known as Spitzer conductivity [Spi13]. NBI is another widely used technique for plasma heating. It uses a beam of high energy neutral atoms to heat and fuel the plasma. The neutral beam can cross the strong magnetic field and is ionized by ions or electrons once it reaches the plasma region. Once ionized, the fast ions will be trapped in the strong magnetic field, losing their energy to the plasma ions and electrons due to electrostatic collisions. Since NBI systems need to generate ions from neutral gas, accelerate them to high energy and neutralize them again these systems are usually very large. Another disadvantage of NBI systems is that it is hard for them to achieve high efficiency, especially at high energy.

RF plasma heating transfers the microwave energy to the thermal kinetic energy of the charged particles based on the cyclotron interaction. It can be separated into two methods according to these different types of charged particles who interact with RF waves: Ion Cyclotron Resonance Heating (ICRH) [SJB+98] and Electron Cyclotron Resonance Heating (ECRH) [FHJ90, LST11, PTZ+13, JAA+13, JGH+13, PAA+15, PIS+08]. In ICRH, the injected RF waves interact mainly directly with ions while for ECRH, it interacts with electrons in the plasma. For efficient transfer of the RF energy for both of these two methods, the RF frequency must be close to the ion or electron cyclotron frequency, because the absorption of energy is most efficient under particle-RF resonant condition, these cyclotron frequencies of ions and electrons are [Che16]

$$\begin{aligned} f_{ci} &= \frac{Z_i e B_0}{2\pi\gamma m_i} \approx \frac{7.63\,\text{MHz} \cdot B_0/\text{T}}{\gamma} \quad (\text{for } {}^2_1\text{D}) \\ f_{ce} &= \frac{e B_0}{2\pi\gamma m_e} \approx \frac{28\,\text{GHz} \cdot B_0/\text{T}}{\gamma} \end{aligned} \tag{1.1}$$

with e: elementary charge, $1.602 \cdot 10^{-19}$ C,
m_e: electron mass, $9.109 \cdot 10^{-31}$ kg,
m_i: ion mass, $3.344 \cdot 10^{-27}$ kg for ${}^2_1$D,
$5.008 \cdot 10^{-27}$ kg for ${}^3_1$T,
Z_i: charge number, 1 for both ${}^2_1$D and ${}^3_1$T,
B_0: magnetic field,
γ: Lorentz factor.

In comparison to ICRH systems, ECRH has to provide a much higher frequency with wavelength in the millimetre region. Therefore, the microwave beam for ECRH can have quasi-optical propagation in the device. This advantage simplifies the design, allows the antenna to be far from the plasma, and reduces the generation of impurities. Once electrons are heated up by ECRH, they can deliver their thermal energy to the plasma ions by Coulomb collision. Due to the high mobility of the electrons in plasmas compared to the ions, this process is relatively fast.

The magnetic fields in modern fusion machines are between 1 T-6 T (e.g. 4.0 T in JET [THP+01] and 5.3 T in ITER [PCE+11]) and the corresponding electron cyclotron frequencies are in the range of 28 GHz-168 GHz. The total heating power requirements for ECRH systems can be up to 24 MW [Thu11]. To limit the costs and increase the reliability of the ECRH systems, the suggested output power for each microwave source unit is around 0.5-2 MW [Sch15]. High power gyrotrons, which can generate up to 1 MW microwaves in the frequency range of 28-170 GHz have been successfully used for ECRH system.

The Institute for Pulsed Power and Microwave Technology (IHM) is supplying 1 MW, 140 GHz gyrotrons for the experimental stellarator Wendelstein 7-X, developing and testing 1 MW, 170 GHz gyrotron for ITER and developing 1 and 2 MW, 240 GHz gyrotrons for the future DEMOnstration Power Plant (DEMO) [JAF+14, JAF+13]. ITER is designed to work with a fusion energy gain factor higher than 10, which means it can generate 500 MW fusion power out of 50 MW input heating power. With this energy gain factor, ITER will be the first fusion device to produce net energy. DEMO will be the prototype commercial reactor with a higher requirement for gyrotrons than ITER. These fusion reactors have high requirements on the efficiency and stability on the ECRH system. With the developing of fusion science, future fusion devices will be bigger and work with a higher magnetic field which will, in turn, have higher requirements on gyrotrons.

1.2 Motivation

The efficiency and stability are two important factors for high power fusion gyrotron operation. Most of the fusion gyrotrons have efficiencies that are below 50 % with Single-stage Depressed Collector (SDC) [Thu15], and some of the gyrotrons come up with Low Frequency Oscillations (LFOs) in the 100 MHz range during their operation [KLP+04, CCM+09, RPK+10].

The quality of the electron beam, which is configured in the Magnetron Injection Gun (MIG), is the key point for efficient and stable gyrotron operation. To fulfil the higher requirement on gyrotron efficiency and stability, more detailed research on these factors, which lead to electron beam quality degradation and possible instabilities in gyrotrons, shall be performed. Emitter surface roughness and emission inhomogeneity are considered to be the two most critical factors.

It is estimated that 5 μm microstructure size of emitter surface roughness will result in a 3.9 % increase of the velocity spread [Edg93]. To simulate the influence of emitter surface on electron beam quality, a simple surface roughness model has been developed by E. Borie for the codes BFCPIC and BFCRAY [BH97]. Simulations with BFCRAY also shows that emitter surface roughness will increase the velocity spread. However, until now, no literature is available to investigate the influence of emitter surface roughness on gyrotron efficiency and stability. In order to simulate the influence, a more accurate surface roughness model is needed.

Another factor that decreases the beam quality is the emission inhomogeneity of the emitter. It has been experimentally tested that the emission inhomogeneity can significantly decrease the gyrotron efficiency [GKV+97] and cause LFOs in the gyrotron [LPS+06b]. The numerical simulation also shows that with typical inhomogeneity of emission, the gyrotron efficiency can be, by a percentage of 2 %-3 %, lower than that with ideally uniform emission [NVB+01]. Yet, a quantitative relation between the emission inhomogeneity and the beam quality is missing. In the meantime, a suitable definition of the emission inhomogeneity and its upper limit for high power gyrotrons shall be clarified.

After long-term operation of the gyrotrons or high voltage breakdown between cathode and anode, the surface roughness and emission inhomogeneity will increase. Those two long-term effects can result in an increase of the number of trapped electrons. The accumulation of trapped electrons in gyrotrons is believed to be one important reason that decreases the gyrotron efficiency, and especially, the stability. However, because a large amount of calculation time needed, the investigation in trapped electrons is quite limited. The behaviour of trapped electrons in the gyrotron is not clear and most of the limited number of research focus on LFOs of the bunched space charge. Another kind of LFO is observed in the short-pulse prototype of the EU 170 GHz 2 MW coaxial cavity gyrotron, which stopped the gyrotron from operating at the nominal point. Still, its mechanism is not clear. In the meantime, damages on the first industrial prototype of the EU 170 GHz 2 MW coaxial cavity gyrotron components are also supposed to be related with Electron Beam Halos (EBHs) generated by trapped electrons. Therefore, this work is carried out in order to study the influence of emitter surface roughness, emission inhomogeneity, and trapped electrons (LFO and EBH) caused by them.

By studying these factors and phenomena, it is possible to predict beam parameters with more accuracy, specify the ways to improve the electron beam quality, gyrotron efficiency and stability.

1.3 Contents and structure of this work

The target of this work is to investigate the influence of surface roughness and emission inhomogeneity on the stability of the electron beam and to find possible improving methods for the future design of gyrotron MIGs. The first part of this work (Chapters 2 and 3) investigates the influence of the quality of the emitter on the quality of electron beam and gyrotron efficiency. LFOs and EBHs are investigated in the second part of this work (Chapters 4, 5 and 6). The flow chart of the structure for this work is shown in Figure 1.1

To investigate the influence of emitter surface roughness on the quality of electron beam and gyrotron efficiency, a surface roughness model is developed for the KIT in-house codes ESRAY and ESPIC [IZJ15] in

Chapter 2. The numerical surface roughness model can be used to simulate the influence of the surface roughness without a costly increase of the mesh number. The influences of surface roughness on beam quality, gyrotron efficiency, and mode competition are calculated using the surface roughness model.

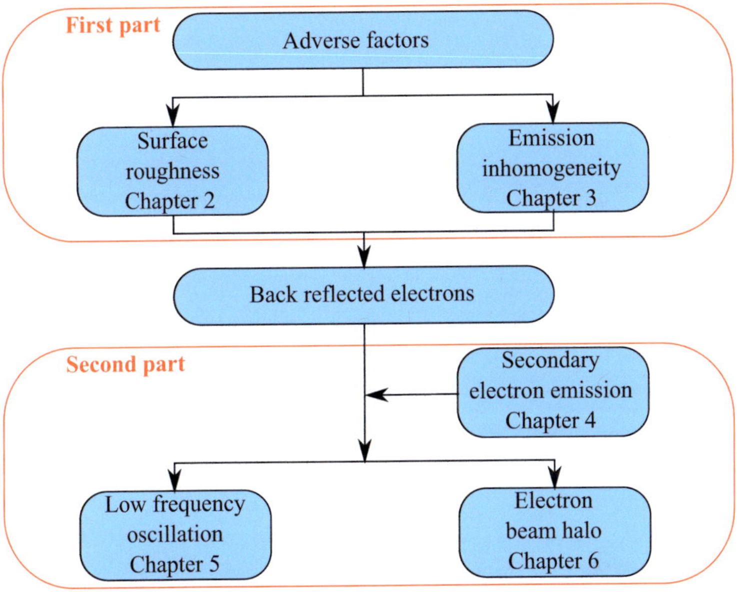

Figure 1.1: Main contents and structure of the thesis.

The emission inhomogeneity information can be evaluated from the measured Current Voltage Characteristics (CVCs). After a review of three existing methods to evaluate the emission inhomogeneity from the CVCs, a new definition of emission inhomogeneity is given in Chapter 3. The new definition is intuitive and stable. The emission inhomogeneities of three different gyrotrons are measured and their influence on two gyrotrons are numerically calculated. At the end of this chapter, general and separate requirements for the emitters are given.

To investigate the influence of back reflected electrons which could be generated by the surface roughness and emission inhomogeneity, a modified Furman secondary emission model is developed for the codes ESRAY and ESPIC in Chapter 4. The new modification gives a more accurate angular distribution of elastically reflected and rediffused electrons. The Secondary Emission Yield (SEY) for molybdenum and tungsten is calculated based on the survey on experimental data.

In Chapter 5, the trapping mechanism of electrons back reflected by the magnetic mirror is described. The space charge LFOs of the short-pulse prototype EU 170 GHz 2 MW CW coaxial cavity gyrotron is presented. The eigenmode oscillation of the coaxial gyrotron is tested and the instability mechanism is explained. In Chapter 6, numerical calculations are used to prove the build-up of the EBHs in the gyrotron MIGs caused by back reflected electrons bombarding the cathode in the vicinity of the emitter. The correlation between observed electron traces at the surface of gyrotron components and the EBH is verified. Three optimization criteria for gyrotron MIGs are presented at the end of this chapter.

1.4 Setup and key components of gyrotron

Gyrotrons are one kind of Vacuum Electron Devices (VEDs), which can generate microwave power at frequencies from several GHz to THz based on the Electron Cyclotron Maser (ECM) instability in a longitudinal magnetic field [Edg93, Thu15, JAF^{+}14, Kup79, Thu03]. The schematic overview of a typical high power gyrotron is illustrated in Figure 1.2 [Thu15, Sch15, RC14, Ber11].

In Figure 1.2, V_{B} is the body/anode voltage, V_{C} is the cathode voltage, B_z is the magnetic field in z direction generated by the superconducting magnet coils. The fundamental working principle of gyrotrons is the following: under the strong electric field between the ring-shaped emitter (cathode) and the anode, electrons are extracted and accelerated. An externally applied magnetic field generated by the superconducting gyrotron magnet guides the electrons towards the cavity, and generates the required gyro-motion of the beam electrons. In the interaction cavity electrons transfer most of their transverse energy of the electron motion to the energy of the electromagnetic field. Within the quasi-optical output system, the microwaves are transferred through the output window. The remaining kinetic energy of electrons is deported at the wall of the single-stage depressed collector. A detailed description of the working mechanism of gyrotrons will be given in the following subsections.

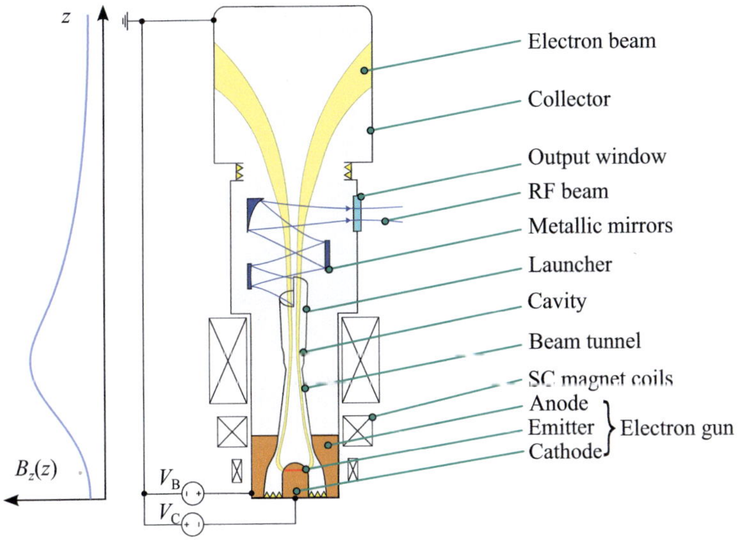

Figure 1.2: 2D sketch of high power gyrotron with the high voltage supply.

1.4.1 Electron beam optics system

The Electron Beam Optics System (EBOS) consists of the MIG and beam tunnel of the gyrotron, which are in front of the cavity. The EBOS generates a hollow electron beam for the cavity with sufficient transverse velocity $v_{\perp}$. In addition, it determines the quality of the electron beam at the cavity. Figure 1.3 shows an axially symmetric cross-section of a typical diode-type EBOS.

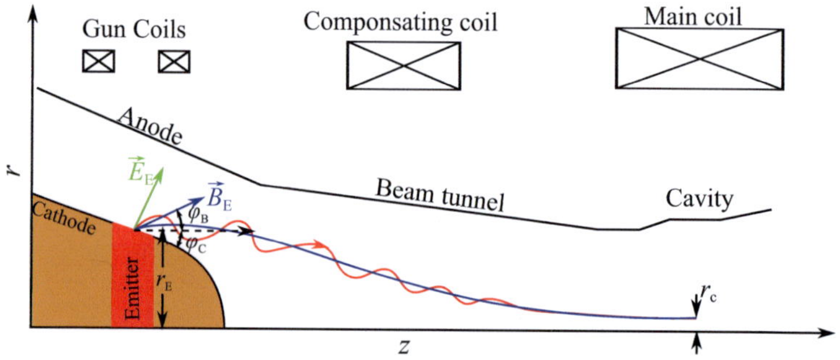

Figure 1.3: Axially symmetric cross-section of a typical diode-type EBOS ($\vec{E}_E$ is the electric field on the emitter surface, $\vec{B}_E$ is the magnetic field on the emitter surface, φ_C is the angle of emitter surface to z-axis, φ_B is the angle of magnetic field to z-axis, r_E is the radius of the emitter and r_c is the radius of the electron beam at the cavity centre).

The setup is surrounded by a superconducting magnet system: The main coils provide the necessary high axial field in the cavity. With a compensating coil, the field produced by the main coils is suppressed in the emitter region. The gun coils provide the possibility for sensitive tuning the desired magnetic field at the starting points of the electron trajectories.

There are two types of MIGs for the EBOS: diode-type MIG and triode-type MIG. In the diode-type MIG (Figure 1.4(a)), the anode and the beam tunnel have the same voltage V_B. In the triode-type MIG (Figure 1.4(b)), an additional modulation anode is installed opposite to the emitter. This additional degree of freedom introduced by the voltage V_{mod} applied to the modulation anode provides the possibility to sensitively adjust the electric field $\vec{E}_{\mathrm{E}}$ in the vicinity of the emitter. The variation of V_{mod} allows a better control over the beam properties, especially the velocity ratio α.

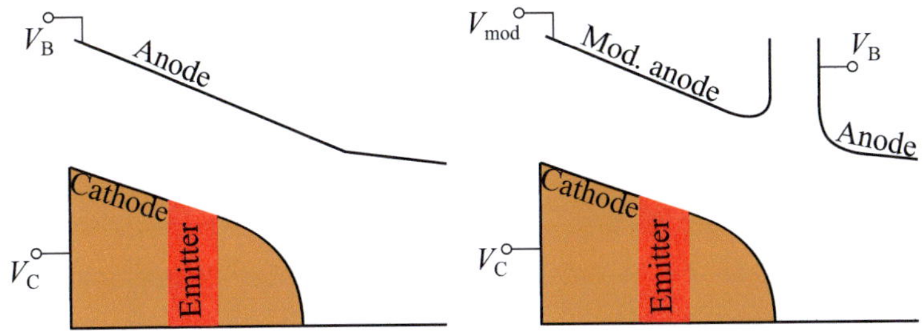

(a) Diode-type MIG

(b) Triode-type MIG

Figure 1.4: Simplified schematics of (a) diode-type and (b) triode-type magnetron injection guns.

The emitter is a key component of the MIG. Usually, impregnated dispenser emitters are used in high power gyrotrons [Thu15]. The impregnated dispenser emitters are made of porous tungsten impregnated with metal oxides. The porous tungsten matrix acts as a reservoir from which the emitting material (metal oxides) can diffuse to the surface, maintain an active layer, and provide a low work function for thermionic emission of electrons [Cro81]. B-type, S-type, and M-type impregnated dispenser emitters are usually used for high power gyrotrons. A detailed description on these three types of emitters will be given in Section 6.2.

With the increase in the electric field on the emitter surface, a thermionic emitter will first work under the space charge limited region and then works in the saturation region or temperature limited region [MG56, TBPS83,

Kho69]. Under space charge limited region, the current density of a parallel-plate diode can be described using the Child-Langmuir equation [Chi11]

$$\begin{aligned} J_{\text{C-L}} &= \frac{4\varepsilon_0}{9}\sqrt{\frac{2e}{m_e}}\frac{V_0^{\frac{3}{2}}}{d^2} \\ &= A_{\text{C-L}}V_0^{\frac{3}{2}} \end{aligned} \tag{1.2}$$

where ε_0 is the free space permittivity, V_0 is the applied voltage, and d is the distance between cathode and anode.

Dispenser emitters are designed to work under temperature limited region. The current density of the thermionic emitter under temperature limited region can be described using the Richardson-Dushman equation with Schottky modification [Cro65]

$$\begin{aligned} J_{\text{S}} &= A_{\text{g}}T^2 e^{-\frac{W-\triangle W}{k_B T}} \\ \triangle W &= \sqrt{\frac{e^3 E}{4\pi\varepsilon_0}} \quad . \end{aligned} \tag{1.3}$$

Here, E is the electric field on the emitter surface, T is the absolute temperature of the emitter, W is the work function of the emitter, k_{B} is the Boltzmann constant, and $A_{\text{g}} = \lambda_{\text{R}} A_0$. A_0 is a universal constant, which equals to $1.20173 \times 10^6 \text{Am}^{-2}\text{K}^{-2}$. λ_{R} is a material-specific correction factor that is typically 0.5. The value $\triangle W$ has to be set to 0, if the Schottky effect is not taken into account which, means the influence of electric field is neglected.

Typical working temperatures of thermionic emitters are around 1000 °C. At such high temperatures and additional strong electric field, electrons can overcome the surface barrier and can escape from the emitter surface. Typical current density on the emitter surface of modern cathodes is in a range from 1 A/cm^2-5 A/cm^2. The beam current is adjusted by the temperature of the emitter and the emitter current source. After the electrons leave the emitter, their movements will be affected by the electric and magnetic field components. The movement is defined by the Lorentz force equation

$$\vec{F} = -e(\vec{E} + \vec{v} \times \vec{B}) \quad . \tag{1.4}$$

Here, $\times$ is the vector product, $\vec{v}$ is the instantaneous velocity, $\vec{E}$ and $\vec{B}$ are external electric and magnetic field.

The first part is the accelerating force of the electric field and the second part is the magnetic force, which focuses the electron gyrating around the guiding magnetic field line. The magnetic force only changes the direction of electrons and it will not change the energy of the electrons. The Larmor radius, which is the distance between the actual spiral trajectory and the guiding centre of the gyrating electron, is given by

$$r_{\mathrm{L}} = \frac{\gamma m_e v_\perp}{eB} \quad . \tag{1.5}$$

Here, $\gamma = 1/\sqrt{1-\beta^2}$ is the Lorentz factor. β is the velocity of electrons in gyrotron, which is normalized to the speed of light c_0

$$\beta = \frac{v}{c_0} \quad . \tag{1.6}$$

At the beginning of the movement, electrons will be accelerated to a high axial velocity component $\beta_{\|}$ and low rotational component $\beta_\perp$ by the anode voltage. Electrons gyrate along the magnetic field lines after the acceleration and come up with a smoothly increasing magnetic field. According to Busch's theorem [Edg93] the relation between beam radius r_{b} and magnetic field is given by

$$\frac{r_\mathrm{b}(z_2)}{r_\mathrm{b}(z_1)} \approx \sqrt{\frac{B(z_1)}{B(z_2)}} \quad . \tag{1.7}$$

The compression ratio of the magnetic field between B_c at the cavity centre and B_E at the emitter surface is defined by

$$b := \frac{B_\mathrm{c}}{B_\mathrm{E}} = (\frac{r_\mathrm{E}}{r_\mathrm{c}})^2 \quad . \tag{1.8}$$

Here, r_E is the radius of the emitter and r_c is the electron beam radius at the centre of the cavity (shown in Figure 1.3). The thickness of the electron beam at the entrance of the cavity $\triangle r_\mathrm{c}$ is determined by the magnetic compression ratio and the width of the emitter, which is given by

$$\triangle r_\mathrm{c} = \frac{r_{\mathrm{E\,max}} - r_{\mathrm{E\,min}}}{b^{1/2}} \quad . \tag{1.9}$$

For efficient interaction between the electron beam and microwaves, the gun design should guarantee lowest values for $\triangle r_\mathrm{c}$ [Ber11]. According to [PPZ+16, PNSA10], $\triangle r_\mathrm{c}$ should be smaller than 1/5 of the wavelength; in this case, the gyrotron efficiency is only decreased slightly. Due to the Larmor rotation of the electrons around the magnetic field line, each electron trajectory has a thickness of $2r_\mathbf{L}$ in the r direction. Therefore, the theoretical minimum $\triangle r_\mathrm{c}$ is given by $2r_\mathbf{L}$.

The application of adiabatic invariants following from Busch's theorem is valid when the variation of the magnetic field along the z-axis is slow [Che74]. As a consequence, the relation between transverse velocity component $\beta_\perp$ and magnetic field B is given by

$$\frac{\beta_\perp(z_2)}{\beta_\perp(z_1)} = \frac{\gamma(z_1)}{\gamma(z_2)}\sqrt{\frac{B(z_2)}{B(z_1)}} \quad . \tag{1.10}$$

Due to the increase of the magnetic flux density from B_E towards B_0 in the axial direction, the electrons will convert their axial velocity $\beta_\parallel$ to transverse velocity $\beta_\perp$. According to [Edg93], $\beta_{\perp c}$ of electrons at the cavity centre depends on the operating parameters in the first approximation as

$$\beta_{\perp c} \approx \frac{b^{1/2}}{\gamma} \frac{|\vec{E}_\mathrm{E} \times \vec{B}_\mathrm{E}|}{c_0 B_\mathrm{E}^2} \quad . \tag{1.11}$$

Equation 1.11 indicates that it is possible to control $\beta_{\perp c}$ by modifying the electric and magnetic field above the emitter surface. A key parameter to characterize the electron beam is the pitch factor α which is defined as the velocity ratio between the transverse velocity component and axial velocity component

$$\alpha := \frac{\beta_\perp}{\beta_\parallel} \quad . \tag{1.12}$$

According to equation 1.10, α will increase along with the increase of magnetic field in z direction. In gyrotrons, α has some spread, because of the surface roughness, emission inhomogeneity, non-uniform magnetic field, and space charge effects. It is possible for some electrons to transfer all of their axial velocity $\beta_\parallel$ to the transverse velocity $\beta_\perp$ when they are travelling in positive z-direction. Electrons with zero $\beta_\parallel$ value will be reflected backwards to the emitter region and will be trapped between the cathode and the cavity. A more detailed discussion on trapped electrons will be given in Chapter 5. The typical value of α at the cavity centre in case of high power gyrotrons is $1.2 < \alpha < 1.5$. The maximum possible α value is limited by the beam spread, which would cause trapped electrons in case of too high α value. The spread of δ_α is defined as

$$\delta_\alpha := \frac{\sigma_\alpha}{\bar{\alpha}} \quad . \tag{1.13}$$

Here, $\bar{\alpha}$ is the average value of α and σ_α is its standard deviation. δ_α is used as the parameter to evaluate the quality of the electron beam. Depending on different cases, the separated velocity spreads $\delta_{\beta_\perp}$ and $\delta_{\beta_\parallel}$ are also used.

The electron kinetic energy is another determining parameter of the electron beam. However, the electron beam itself constitutes a significant amount of negative charge inside the gyrotron body. In combination with the external potential of the metallic geometry itself, this imposes a repelling potential on electrons. The consequence is a decrease ("depression") of effective beam voltage V_b with respect to the technical accelerating voltage V_{acc}. The final beam voltage is given by

$$V_b = V_{acc} - \triangle V \quad . \tag{1.14}$$

Here, $\triangle V$ is the depression voltage. For conventional hollow cylindrical cavity gyrotrons, the voltage depression of the thin azimuthally symmetric annular electron beam is given by [DK81]

$$\triangle V = \frac{I_b}{2\pi\epsilon_0 v_{||}} \ln(\frac{r_w}{r_b}) \quad . \tag{1.15}$$

Here, r_w is the radius of the metallic wall, r_b is the electron beam radius and I_b is the beam current. Taking the nominal parameter of a high power gyrotron for example, V_{acc} = 79.5 kV, I_b = 40 A, the wall radius at the centre of the cavity r_w = 19.2 mm, r_c = 9.45 mm. The depression voltage is approximately 6.93 kV and the average electron energy at the cavity centre will be E_{kin} = 72.53 keV. The energy spread of the electron kinetic energy $\delta_{E_{kin}}$ at the cavity centre is another important parameter to evaluate the quality of the electron beam. The gyrotron efficiency is very sensitive to the kinetic energy spread.

1.4.2 Cavity

After the electrons have reached the cavity region along the magnetic field line the cyclotron interaction will take place. The main part of the cavity is usually a straight circular cylinder. The centre is located at the maximum point of the magnetic field. The electromagnetic field in cylindrical waveguides is described by Transverse Electric (TE) and Transverse Magnetic (TM) modes [Poz09]. For TM modes, the transverse field components E_φ and E_r vanish in a long cavity with a resonance close to cut-off. Consequently, no efficient energy exchange with the transverse velocity component of the electrons is possible. Stable and efficient gyrotron operation is only possible with TE modes, because their transverse field components do not vanish [Ber11].

In the electric field of TE modes, electrons can be both accelerated and decelerated during their gyration. Figure 1.5 is the sketch of the interaction between the electric field and electrons. Assuming two typical positions of the gyrating electron in the electric field: one in the decelerating field (left), one in the accelerating field (right).

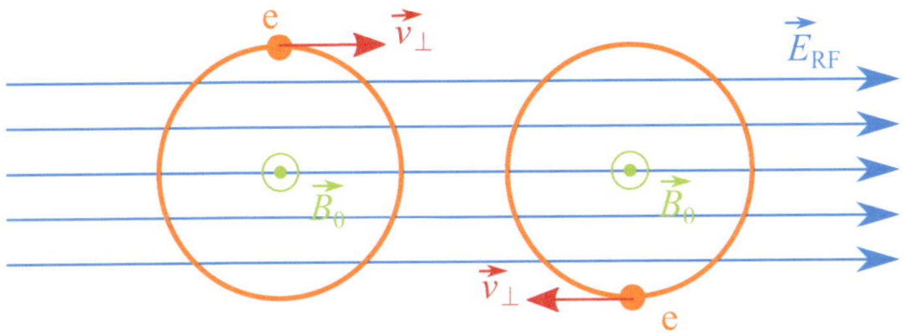

Figure 1.5: Interaction between electrons and electric field in the magnetic field ($\vec{E}_{\mathrm{RF}}$ is the electric field of the microwaves).

When the electron is in the left position, it will be decelerated by the electric field of the microwaves and lose energy to the field. With the decrease of velocity, the Lorentz factor γ will also decrease. According to equation 1.1, the angular frequency of the electron $\Omega_{\mathrm{c}} = 2\pi \cdot f_{\mathrm{ce}}$ will increase due to the relativistic effect. At the right position, the electron is accelerated by the electric field and Ω_{c} will decrease. If the angular frequency of the microwaves $\omega_{\mathrm{RF}} > \Omega_{\mathrm{c}}$, the increase of Ω_{c} for the electron in left position will slow down the phase difference between the electric field and the electron. In this case, it can stay longer in the phase position where it loses energy. In the right position, the electron velocity is increased by the electric field and therefore it will remain for a shorter time in the phase position where it gains energy. In summary, the electrons accumulate in a particular relative phase position where they transfer energy to the RF-field. This focusing towards one phase is called "bunching". If $\omega_{\mathrm{RF}} < \Omega_{\mathrm{c}}$, electrons will accumulate in the phase position where they gain energy from the RF-field. To enable a beneficial energy exchange, the oscillation frequency ω_{RF} has to be slightly higher than the angular velocity of the electrons

$$\omega_{\mathrm{RF}} \gtrsim \Omega_{\mathrm{c}} \quad . \tag{1.16}$$

The propagation of microwaves in the cavity can be characterized by the dispersion relation:

$$k_0^2 = k_\perp^2 + k_\parallel^2 \tag{1.17}$$

$$k_\perp = \frac{\chi'_{m,n}}{r_{\mathrm{w}}} \quad . \tag{1.18}$$

Here, $k_0 = \frac{\omega_{\mathrm{RF}}}{c_0} = \frac{2\pi}{\lambda}$ is the free space wave number, r_{w} is the cavity radius, and $\chi'_{m,n}$ is the nth zero in the derivative of the mth order first kind Bessel function. It shall be pointed out that only the modes with $k_\parallel > 0$ can propagate through the cavity. $k_\parallel = 0$ gives the cut-off frequency of the mode as

$$f_{\text{cut-off}} = \frac{c_0 \chi'_{m,n}}{2\pi r_{\mathrm{w}}} \quad . \tag{1.19}$$

Considering the relative motion $v_{\|}t$ of the electrons to the RF-field and phase of the electric field $(\omega_{\mathrm{RF}} - k_{\|}z)v_{\|}$ at the position of the electron, the general requirement of the resonance condition is [Pet99, Chu04]

$$\omega_{\mathrm{RF}} - k_{\|}v_{\|} \cong s \cdot \Omega_{\mathrm{c}}, \quad s = 1, 2, ... \quad . \tag{1.20}$$

Here, s is the harmonic number, $s = 1$ corresponds to the fundamental cyclotron frequency. Gyrotrons can also operate at higher harmonics, which decrease the requirement for the magnetic field by a factor of s.

The Brillouin diagram [Thu15, Chu04], also called dispersion diagram, shows the resonance region between the electromagnetic mode following equation 1.17, and the electron beam following equation 1.20. As shown in Figure 1.6, the wave angular frequency ω is plotted as a function of the axial wave number $k_{\|}$. The hyperbolas show dispersion curves of two allowed propagating TE modes in the cavity and the two red lines give the electron beam parameters. The intersection of the dispersion curve with the beam line gives the resonance points between the electron beam and electromagnetic wave mode.

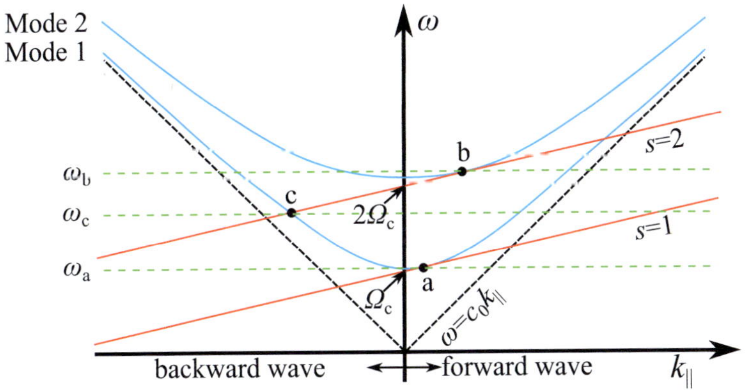

Figure 1.6: Dispersion diagram of two waveguide modes, electron beam line, resonance frequency and speed-of-light-lines (dashed).

Point (a) demonstrates the fundamental gyrotron operating point where the beam line is tangent to the dispersion curve. $k_{\parallel}$ is small and the resonance frequency ω_{a} is near the cut-off frequency of mode 1. Point (b) indicates the resonance point of forward mode 2 and second harmonic of the cyclotron frequency. A backward wave whose corresponding $k_{\parallel}$ is negative can also be generated as is shown in point (c). The backward wave interacts with the second harmonic of the electron cyclotron frequency. The resonance frequency ω_{c}, in this case is lower than $2\Omega_{\mathrm{c}}$. It shall be notified that one beam line can have several interaction points with different modes; more complicated models shall be used for gyrotron beam-wave interaction [Jel00, Ker96].

$\eta_{\perp}$ is the efficiency of the energy transfer from the transverse electron power $P_{\perp}$ of the gyrating electrons to the generated RF-power in the cavity P_{elec}.

$$\eta_{\perp} = \frac{P_{\mathrm{elec}}}{P_{\perp}} \tag{1.21}$$

The electrons can not transfer all their gyrating power to the electromagnetic wave. The theoretical upper limit of $\eta_{\perp}$ is approximately 72% [KDST85], but due to the influence of velocity, energy, and guiding centre spread, the real value is always lower. The efficiency of the total energy removal from the electrons is given by

$$\eta_{\mathrm{elec}} = \frac{\alpha^2}{1+\alpha^2}\eta_{\perp} \quad . \tag{1.22}$$

1.4.3 Quasi-optical output system

After the beam-wave interaction in the cavity, the generated microwaves will propagate through the quasi-optical mode converter. The use of a mode converter has two purposes: separate the electromagnetic wave from the spent electron beam and convert the high order complex wave mode into a paraxial beam for low loss transmission. The quasi-optical output system can be separated into two parts: launcher and reflectors (mirrors), as shown in Figure 1.2. The launcher, which is also called antenna, is a waveguide with specific perturbations of the inner wall. A defined mode mixture process happens along the launcher and a Gaussian beam (TEM_{00}) is generated at the end. The generated Gaussian beam is focused and reflected out of the gyrotron through the output window by the reflectors. Detailed calculations and optimization process for quasi-optical output system can be found in [Jin07, Fla12]

1.4.4 Depressed collector

After the electron beam is separated from the microwave beam, the electrons will travel to the end of the gyrotron and will be collected by the collector. The use of depressed collector in gyrotrons results in a significant increase in gyrotron efficiency [PBD+99], since it collects part of the remaining kinetic energy of the electrons after they have transferred a part of their energy to microwave radiation. The collector efficiency is

$$\eta_{\mathrm{col}} = \frac{V_{\mathrm{acc}}}{V_{\mathrm{acc}} - V_{\mathrm{col}}} \quad . \tag{1.23}$$

Here, V_{col} is the collector depression voltage. The maximal achievable V_{col} is influenced by the minimum electron velocity after the interaction. Due to the velocity spread, parts of the electrons could be reflected under high collector voltage. Therefore, a moderate value will be chosen for V_{col}. The depression voltage of the collector not only increases the gyrotron efficiency, but also decreases the requirements for the cooling system and increases the lifetime of the collector.

1.5 Tools used in gyrotron simulation

For the design and calculation of different components for gyrotrons, different KIT in-house codes are used in IHM. The gyrotron components and the specific codes used are listed in Table 1.1.

Table 1.1: Simulation codes used for different gyrotron components.

Component	**Code**	**Chapters**	**References**
EBOS	ESRAY	2	[IZJ15]
	ESPIC	5, 6	
	ARIADNE	3	[PV04]
Cavity	EURIDICE	2, 3	[APIV12]
Quasi-optical mode converter	TWLO		[JTP+09]
Collector	ESRAY		[IZJ15]

In this work, the code ESRAY is used in the calculation of the influence of surface roughness on electron beam quality generated by the EBOS system (Chapter 2). During the simulation of emission inhomogeneity on the electron beam quality the three-dimensional (3D) code ARIADNE is used since emission inhomogeneity is a 3D problem (Chapter 3). For the calculation of beam-wave interaction, the code EURIDICE is employed (Chapter 2 and Chapter 3). The code ESPIC is used for the PIC simulation of dynamic phenomena caused by surface roughness and emission inhomogeneity in Chapter 5 and Chapter 6. More detailed introduction of the codes will be given in the following parts.

1.5.1 Simulation of electron beam optics system

ESRAY [IZJ15] and ARIADNE [PV04] are self-consistent electrostatic codes, which calculate the electron beam trajectories according to the Finite Difference Method (FDM) and the Finite Element Method (FEM), respectively. ESRAY has been developed for 2D simulation while ARIADNE was initially created for solving 3D problems. ESPIC is a quasi-static Particle In Cell (PIC) code. It is an extension of the electrostatic code ESRAY, which can simulate slowly varying effects in the gyrotron MIG. The classical PIC approach is used for the particle handling. These codes are used in this work for separate simulations according to their different features.

ESRAY

The core of the ESRAY beam optics code is a finite-difference potential solver based on the multi-grid method, operating on a two-dimensional, non-orthogonal (but structured) simulation grid in cylindrical coordinates. This concept offers less flexibility compared to typical meshes used in finite element codes, but the multi-grid solver is relatively fast and has a moderate demand on memory usage. Figure 1.7 shows details of such a simulation mesh for the case of an exemplary gyrotron.

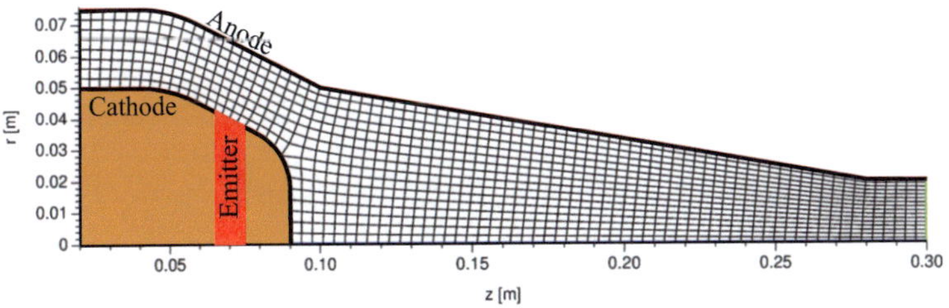

Figure 1.7: ESRAY simulation grid showing the cathode, anode and body of an exemplary gyrotron.

For particle pushing, the energy-conserving leap-frog scheme according to Buneman [Bun67] is used. Both schemes together provide second order accuracy, as long as the shape of the grid cells does not differ too much from the ideal rectangular shape. The structured mesh allows a very fast localization of macro-particles in the non-orthogonal mesh and the interpolation of forces to the particle position can also be implemented in a CPU-efficient way. Several emission models are available, including thermionic emission, which takes the influence of the electric field strength at the emitter surface into account (Schottky effect [Cro65]). During a simulation run, the calculation of the full electron beam trajectories are alternated with the calculation of the self-induced electric field until a self-consistent solution is reached.

The ESRAY code package consists of the following three main modules that have to be processed in sequential order:

Gridgen:

Generates the simulation mesh from a geometrical description of border sections stored in an input file for geometry.

Maggen:

Calculates the magnetic flux density on all mesh nodes according to the geometries, winding numbers and applied currents of the solenoids of the magnetic system. The required input information is stored in an input file for coils.

ESRAY:

Performs the self-consistent trajectory simulation using the provided simulation mesh and the magnetic field profiles of the individual solenoids. Fig. 1.8 gives an overview of the structure of the main iteration loop of the simulation program. The target of the loop is to obtain a self-consistent solution. The calculation of the full electron beam trajectories is alternated with the calculation of the self-induced electric field (typically takes 10 - 25 iterations). Depending on the computer hardware, mesh size, and the requested accuracy, one ESRAY run typically takes several seconds to several minutes.

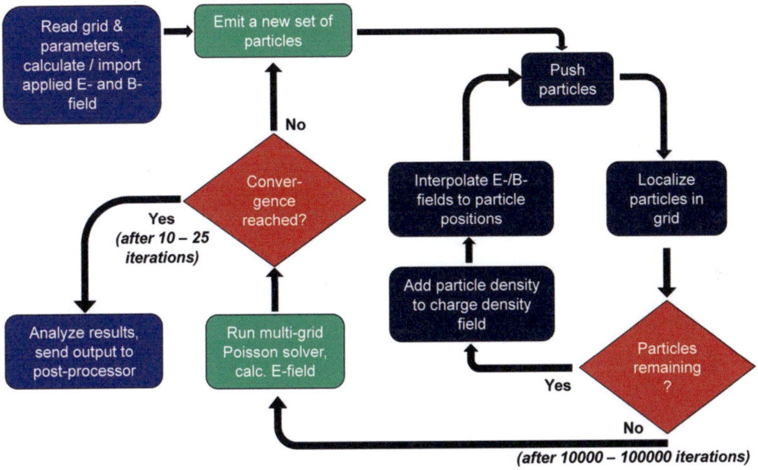

Figure 1.8: Organization of the main iteration loop of ESRAY.

ESPIC

ESPIC is a slightly modified version of ESRAY. It is used to simulate the slowly varying effects in the gyrotron electron gun and the beam tunnel. This code does the particle handling in the classical PIC approach using a 2D multi-grid Poisson-solver for the calculation of the electric fields; a full handling of Maxwell's equation is not yet implemented. With the surface roughness model and secondary electron emission model, investigations on LFOs and secondary electron beam halo is carried out in this work.

ARIADNE

ARIADNE is a parallel 2D and 3D trajectory code for gyrotron electron beam simulation. Similar to ESRAY, ARIADNE is also a self-consistent electrostatic code. However, the finite element method is used for the solution of the Poisson equation in contrast to ESRAY. Finite element curvilinear cubic, tetrahedral meshes, and curvilinear quadrilateral meshes can be generated by ARIADNE. Figure 1.9 shows details of a finite element curvilinear cubic mesh for the case of an exemplary gyrotron.

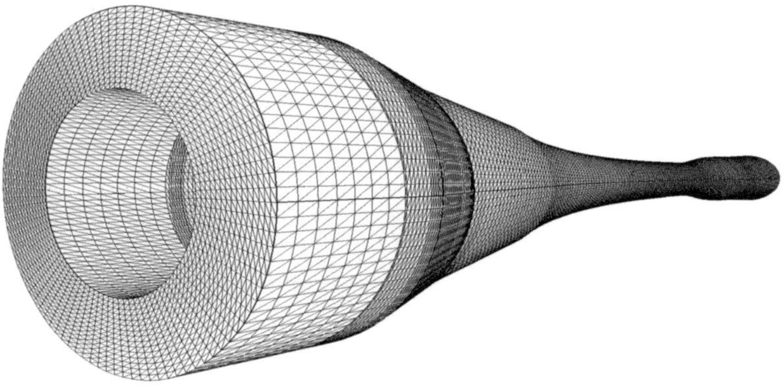

Figure 1.9: Finite element curvilinear cubic mesh used in ARIADNE simulation showing an exemplary gyrotron.

For a conventional gyrotron ARIADNE can be used to study effects associated with deviations from cylindrical symmetry, such as those produced by non-uniform electron emission from the cathode, misalignments between the mechanical and the magnetic axes, etc. In addition, it can handle fully 3D geometries (with dependence on the polar angle), as is the case of a sheet beam for a quasi-optical gyrotron.

The basic steps in the simulation of a gyrotron electron beam using ARIADNE are the following: (i) The user introduces the 2D/3D geometry and the boundary conditions in the database of the program; (ii) The mesh generator of the code, based on mapping technique, automatically creates a finite element curvilinear cubic or tetrahedral mesh for 3D geometries and curvilinear quadrilateral mesh for 2D geometries; (iii) The solver subroutine, using the finite element method, initially solves the Laplace equation and calculates the potential on the mesh nodes due to the boundary conditions; (iv) The particle pusher subroutine, which is based on Runge-Kutta method, calculates the first approximation of the beam electrons' trajectories; (v) Self-consistency is obtained by successive application of the Poisson solver (to obtain the electric field in the charge density of the previous iteration) and the particle pusher (to obtain the charge density due to the fields of the previous iteration), until convergence is achieved to the desired accuracy. More detailed introductions to ARIADNE can be found in [PV04].

1.5.2 Simulation of beam-wave interaction in the cavity

EURIDICE is a fast code-package for gyrotron interaction simulations and cavity design. It comprises codes for operating mode selection, calculation of the field profile of TE modes in the cold cavity, calculation of the modes' starting currents, self-consistent calculation of the field profile in steady-state single-mode operation, and time-dependent multi-mode simulation of the beam-field interaction. The calculation algorithm of the time-dependent multi-mode simulation part used in this work is shown in Figure 1.10.

EURIDICE can treat an arbitrary number of TE modes. Simultaneous interaction of different modes at different cyclotron harmonics is permitted. The field profiles of the modes can be those of the cold cavity or others, user-defined. The output power and frequency pulling of the modes are obtained as functions of time. The beam parameters (i.e., voltage, current, α) can vary over time according to the adiabatic approximation and the Schottky formula for temperature-limited emission (equation 1.3), or according to user defined tables (obtained from electron gun simulation codes, or from experiment). The variation of the parameters can be either linear or stepwise. In this way, start-up simulations are feasible. The time-dependent part is parallelized, allowing fast multi-mode simulations. Ohmic losses, axial variation of the magnetostatic field, and spreads in the electron beam parameters are also considered. Detailed information about EURIDICE can be found in [APIV12].

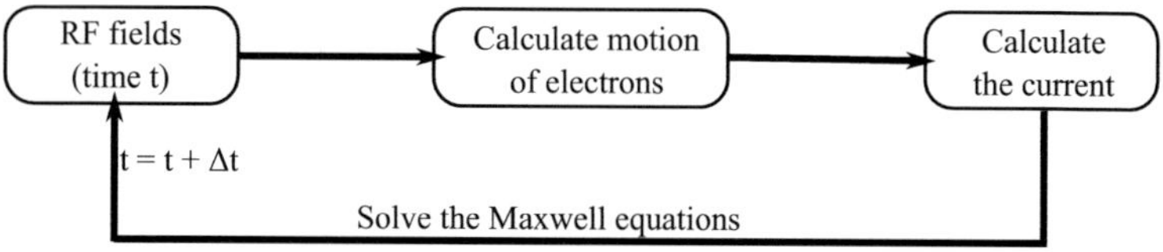

Figure 1.10: Numerical algorithm of the main iteration loop for the time-dependent part of EURIDICE.

1.6 Gyrotron models used in this work

Different gyrotron models are used in the calculations and experiments of this work. The information for these gyrotrons and the corresponding chapters they are used in are shown in Table 1.2.

In Table 1.2, the first column is the abbreviation of the gyrotrons. The secondary column is the type of the emitter. The third and fourth columns give the work function and typical working temperature of the emitters. These work functions are provided by the manufacturer while the temperatures are calculated by the fitting method discussed in Chapter 3.

Table 1.2: Gyrotrons models used in the simulation and experiment of this work.

Gyrotron	Emitter	W [eV]	T [K]	Simulation / Measurement		Chapter	Figure
1 MW, 170 GHz	B-type	2.1	1310	MIG-1	Surface roughness	2	6.1
					Electron beam halo	6	
				MIG-2	Emission inhomogeneity	3	6.12
					Electron beam halo	6	
2 MW, 170 GHz	B-type	2.1	1308	Short-pulse prototype	Emission inhomogeneity	3	5.2
					Low frequency oscillation	5	5.6
				Industrial prototype	Electron beam halo	6	6.6
15 kW, 28 GHz	M-type	1.85	1173	Emission inhomogeneity		3	6.16
				Electron beam halo		6	

The simulations and experiments where these gyrotrons are used are given in column five. The last two columns give the corresponding chapters and the schematic plots for these gyrotrons.

In Chapter 2, two MIG designs for the EU 1 MW, 170 GHz ITER gyrotron [PIS+08] are used to calculate the influence of surface roughness on electron beam quality and gyrotron efficiency. MIG-1 is the initial design of the MIG, while MIG-2 is the optimization of MIG-1, which can suppress the generation of EBH in the gyrotron MIG. In Chapter 3, the emission inhomogeneity of MIG-2 for the EU 1 MW, 170 GHz ITER gyrotron, the KIT 15 kW, 28 GHz gyrotron for materials processing [MIP+13], and the short-pulse prototype of the EU 2 MW, 170 GHz coaxial cavity gyrotron [PDD+04, RPK+10] have been measured and analysed. Chapters 5 and 6 focus on the two phenomena (LFOs and EBHs) caused by the trapped electrons. The short-pulse prototype of the EU 2 MW, 170 GHz coaxial cavity gyrotron is used for the investigation on LFOs since such a phenomenon has been observed during the experiment of this gyrotron. The first industrial prototype of the EU 2 MW, 170 GHz coaxial cavity gyrotron and these two MIGs for the EU 1 MW, 170 GHz ITER gyrotron are also used for the calculation of EBHs.

2 Surface roughness model and its influence on gyrotron

This chapter surveys the influence of the emitter surface roughness on the operation of high power gyrotrons. After a review on the influence of surface roughness at the beginning of this chapter, a surface roughness model is derived for the trajectory code ESRAY and the quasi-static Particle In Cell (PIC) code ESPIC [IZJ15]. Then, the influence of surface roughness on the beam quality and gyrotron efficiency is quantitatively studied. The code ESRAY was used for MIG simulation. The code EURIDICE [APIV12], which can import the electron beam from ESRAY, was used for the study of the influence on gyrotron efficiency and mode competition. Phenomena, such as the generation of magnetically trapped electrons, are predicted at a high level of roughness for the first time. Detailed discussions on trapped electrons will be presented in Chapter 5 and Chapter 6.

2.1 Introduction

The quality of the electron beam is one of the key points for efficient and stable gyrotron operation. A proper electron beam will have low spread in velocity, energy, guiding centre and current density. The influence of energy and velocity spread on gyrotron efficiency of a gyrotron has been numerically calculated using an electron beam with different given spreads [ZZM06]. The following relation was derived

$$\eta_{\mathrm{elec}} = \eta_{\mathrm{elec,max}}[1 - K_{E_{\mathrm{kin}}}(\delta E_{\mathrm{kin}})^2 - K_{v_\perp}(\delta v_\perp)^2] \quad . \tag{2.1}$$

Here, η_{elec} is the interaction efficiency and $\eta_{\mathrm{elec,max}}$ is the maximum efficiency in the case of no spreads, $K_{E_{\mathrm{kin}}}$ is the coefficient of energy spread, and $K_{v_\perp}$ is the coefficient of velocity spread. The efficiency will decrease with the increase of energy and transverse velocity spread. Therefore, the general purpose of the conventional gun design is to generate an ideal beam with low spread. However, in reality, several factors have a negative impact on the beam quality [Tsi01, Tsi72, Edg95, ZZM06, GGM$^+$10, LAL96, Lau87], which are

1. the emitter surface roughness,
2. the current emission inhomogeneity at the emitter surface,
3. the spread of the electron initial thermal velocities,
4. the asymmetrical electric and magnetic fields in the MIG due to the misalignment,
5. the low frequency oscillation of trapped electrons between the magnetic mirror at the entrance of the cavity and the cathode.

This thesis mainly focuses on factors (1), (2) and (5). Detailed discussion on the influence of (4) is presented in [ZT96, DP97, ISN$^+$98]. The influence of (3) can be numerically calculated by imparting initial thermal velocities to the electrons when they are emitted from the cathode surface. To achieve the condition of thermionic emission, the cathode is heated to a high temperature, and electrons are emitted from the emitter surface with initial thermal energy. The initial energy has a Maxwell-Boltzmann (M-B) distribution [RC70], which is given by

$$f_{M-B} = \frac{1}{k_B T} e^{-\frac{E_{\text{kin}}}{k_B T}} \quad . \tag{2.2}$$

Here, E_{kin} is the kinetic energy of the electron, k_B is the Boltzmann constant, and T is the absolute temperature at the emitter surface. For a typical emitter temperature $T = 1000$ °C, the average initial average thermal energy is $k_B T \approx 0.11$ eV.

Due to the machining, bombardment of positive ions, and evaporation of barium from the porous tungsten surface, roughness appears at the emitter surface. A similar approach is followed for the estimation of the influence of emitter surface roughness [Tsi72]. The theoretical estimation uses the initial velocity to investigate the influence of surface roughness; the velocity spread caused by the surface roughness can be estimated by

$$\delta v_{\perp} = 1.6 \sqrt{(1 + \frac{\pi^2}{4} \tan^2 \varphi_{\text{BC}}) \frac{r_0}{h}} \quad . \tag{2.3}$$

Here, $\delta v_{\perp}$ is the relative spread of the perpendicular electron velocities, r_0 is the average roughness height, h is the height of the first vertex of the electron trajectory above the cathode, and $\varphi_{\text{BC}} = \varphi_{\text{B}} + \varphi_{\text{C}}$ is the inclination angle of the magnetic field with respect to the emitter surface, as shown in Figure 1.3.

Combining equations 2.1and 2.3 one gets

$$\eta = \eta_{\max}[1 - K_{U_B}(\delta U)^2 - 2.56 K_{v_{\perp}}(1 + \frac{\pi^2}{4} \tan^2 \varphi) \frac{r_0}{h}] \quad . \tag{2.4}$$

Equations 2.3 and 2.4 provide the approximate velocity spread of the electrons and its influence on the gyrotron efficiency. In addition to the decreasing gyrotron efficiency due to the emitter surface roughness, several other phenomena in the gyrotron can be caused, such as back reflected electrons resulting in secondary electron emission and low frequency oscillations.

In principle, the influence of the emitter surface roughness can be simulated by 3-D codes using physical roughness models. However, a huge mesh cell number is necessary for the modelling of surface roughness, since the microstructure size on the emitter surface is of the order of several microns. The computer memory needed and the calculation time is unacceptable. Therefore, instead of modelling the realistic emitter surface roughness, a simple approach is derived in this thesis.

2.2 Surface roughness model

Similar to [Tsi72], to study the influence of surface roughness on the beam quality using a trajectory code, an initial velocity distribution of the emitted electrons is considered. The definition of initial velocity is based on the extension of the roughness model only by hemispherical microstructures of varying size to a more sophisticated structure, which consists of an ensemble of cylindrical, conical, and hemispherical microstructures, as shown in Figure 2.1. The dashed vertical lines in Figure 2.1 show for each microstructure the axis of rotational symmetry. The emitter surface roughness model is developed as one built-in function of the electrostatic code ESRAY [ZIPJ13].

E_0 is the electric field on the emitter surface in ideal case whereas the parameter r_0 is the microstructure size. The influence of each type of microstructure on the emitted electrons is numerically studied for each type of microstructure individually. A simple capacitor model with two parallel plates whose distance is several times larger than the microstructure size r_0 is used for the calculation of each type of microstructure. The velocity distribution of the electrons recorded on the anode is used for the definition of the initial velocity of the electrons emitted from the emitter ring for electron beam simulation. By this approach, the initial velocity of the emitted particles is calculated versus the applied electric field, the microstructure size, and the relative position of the emission. Figure 2.2 shows the 2D axial-symmetrical trajectories of these electrons for all kinds of surface defects used in the capacitor model. Electrons from the plane area are also taken into account since the microstructure will also influence their trajectories. (I) represents different parts of the microstructures and (II) corresponds to parts of the plane area. Parts of the trajectories of emitted electrons cross the axis of symmetry $r = 0$ for the inverse microstructures.

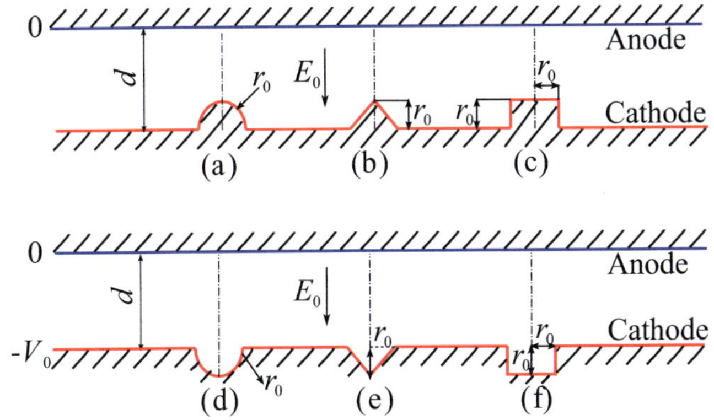

Figure 2.1: Microstructures used in the surface roughness model (a: hemispherical bump, b: conical bump, c: cylindrical bump, d: inverse hemispherical bump, e: inverse conical bump, f: inverse cylindrical bump).

According to [Tsi72], the influence of the microstructure on the quality of the electron beam can be calculated by using the initial velocity component, which is perpendicular to the normal vector of the smooth emitter surface. Therefore, the perpendicular velocity component on the anode of the capacitor model is pre-defined as the initial velocity $\beta_{\text{ini},\perp}$ of the electrons in the MIG simulation codes ESRAY and ESPIC, as shown in Figure 2.3.

Using the initial velocity model will bring artificial energy to the system, which does not exist in reality. Therefore, while simulating the influence of surface roughness, the initial velocity should be as small as possible.

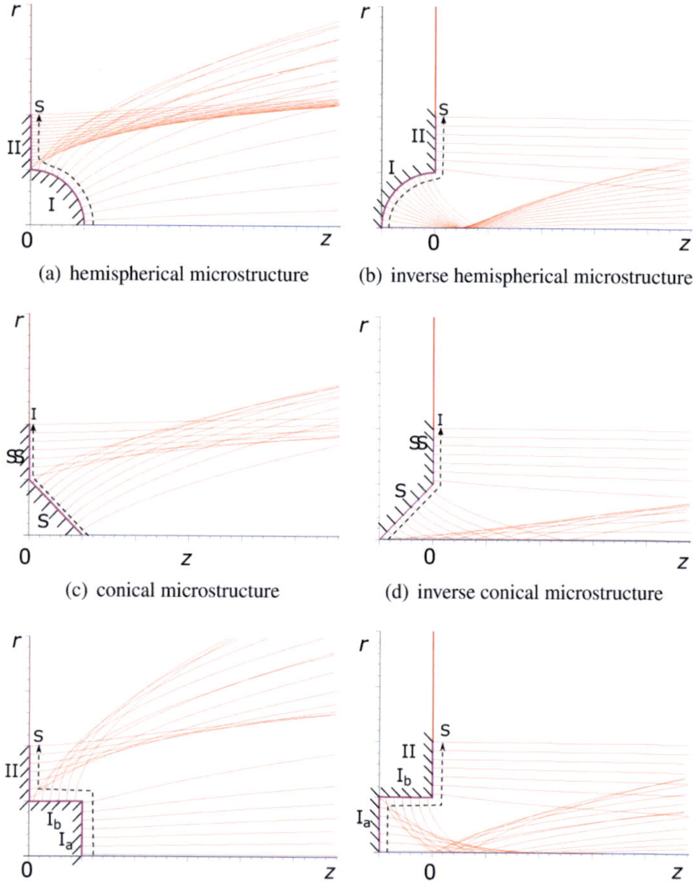

(a) hemispherical microstructure (b) inverse hemispherical microstructure

(c) conical microstructure (d) inverse conical microstructure

(e) cylindrical microstructure (f) inverse cylindrical microstructure

Figure 2.2: Electron trajectories for different kinds of microstructures (region I). Parts of the plane area are also taken into account (region II). $E_0 = 30\,\mathrm{kV/cm}$, $r_0 = 5\,\mu\mathrm{m}$.

Neglecting the initial parallel velocity also results in this model creating less artificial energy, since $\beta_{\mathrm{ini},\perp}$ is several times smaller than $\beta_{\mathrm{ini},\parallel}$ at the distance (several times r_0) where the velocity information is collected, as is shown in Figure 2.4.

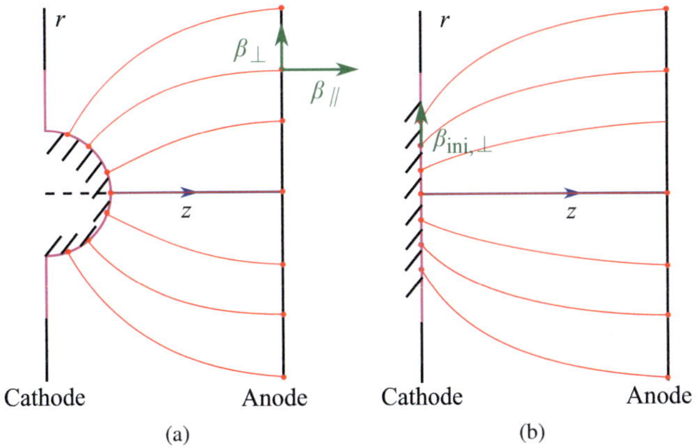

Figure 2.3: Schematic of (a) the electrons emitted from the hemispherical microstructure and (b) the corresponding initial velocity model used in ESRAY.

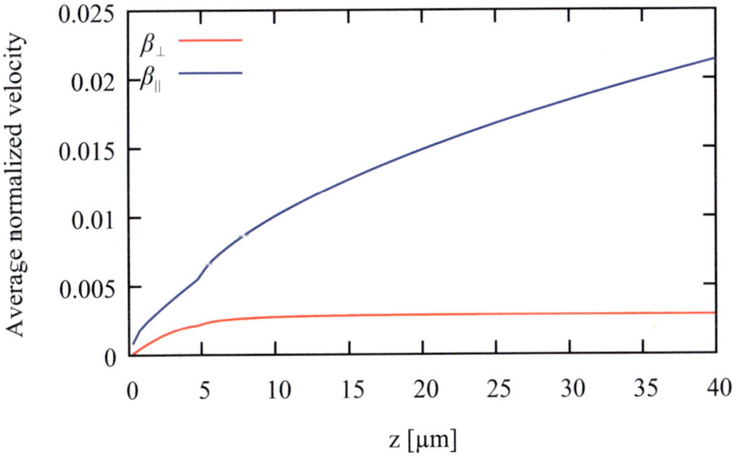

Figure 2.4: Average velocity components along z-axis of a hemispherical microstructure.

According to [BLC$^+$08], present cathodes generally have surface roughness in the range of 2.5 µm-5 µm. Hence, a hemispherical microstructure of 5 µm size with applied electric field $E_0 = 30\,\text{kV/cm}$ is used for the calculation. The velocity information at 7 times r_0 is collected and used as the initial perpendicular velocity $\beta_{\text{ini},\perp}$. It is shown that the normalized velocity component $\beta_{\|}$ is approximately 7 times larger than $\beta_{\perp}$. The discontinuity of $\beta_{\perp}$ and $\beta_{\|}$ at z =5 µm comes from the fact that the microstructure ends at 5 µm.

The initial perpendicular velocity of the electron versus the position of emission for the six types of microstructures is shown in Figure 2.5.

In Figure 2.5, the starting position S is the route from the intersecting point of the microstructure outline with the z-axis to the end of the flat area (as shown in Figure 2.2). In all cases, inverse microstructures have smaller initial velocity and therefore a smaller influence, due to the lower electric field on the surface of the microstructure. The average value of initial normalized velocity $\beta_{\text{ini},\perp}$ caused by the surface roughness is about 0.002 corresponding to 1.0 eV of initial electron energy. This value is significantly higher than the typical energy corresponding to the initial thermal velocity (see Section 2.1) at 1000 °C.

Figure 2.5 provides the needed information at one typical condition ($E_0 = 30\,\text{kV/cm}$, $r_0 = 5\,\mu\text{m}$). In reality, the emitter works under a range of E_0 and the size of the emitter surface roughness is also a distribution. Therefore, it is necessary to find the dependency of $\beta_{\text{ini},\perp}$ on the microstructure size r_0 and the applied electric field strength E_0. To find that relation, an analytical calculation is made using the hemispherical microstructure as an example. According to [FEV03], the potential distribution near the hemisphere is

$$\varphi = r\cos\theta E_0\left(1 - \left(\frac{r_0}{r}\right)^3\right) \quad . \tag{2.5}$$

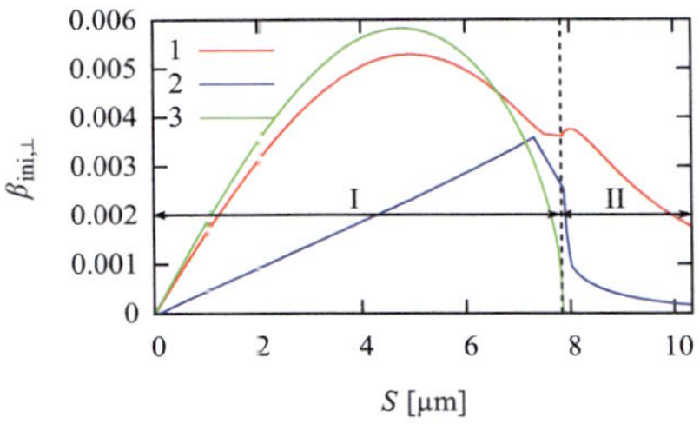

(a) Hemispherical and inverse hemispherical microstructure

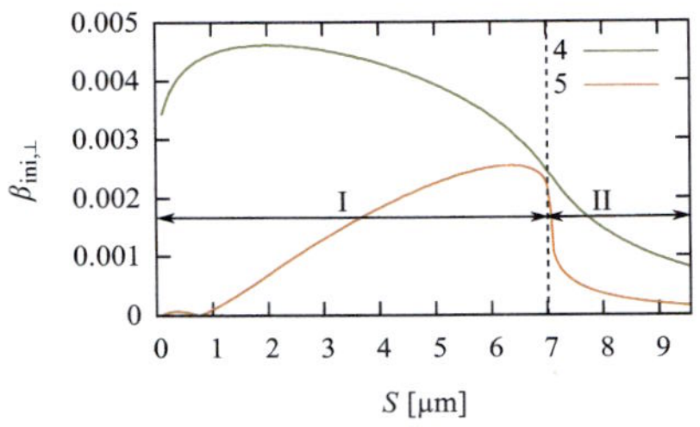

(b) Cone-shaped and inverse hemispherical microstructure

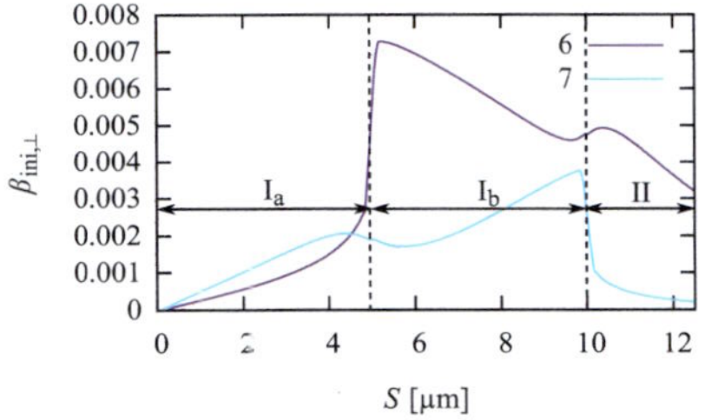

(c) Cylindrical and inverse cylindrical microstructure

1: Hemispherical bump.
2: Inverse hemispherical bump.
3: Hemispherical bump (theoretical).
4: Cone-shaped bump.
5: Inverse cone-shaped bump.
6: Cylindrical bump.
7: Inverse cylindrical bump.

Figure 2.5: Relations between the starting position S of the electrons from the microstructure and the initial perpendicular velocity $\beta_{\mathrm{ini},\perp}$ for $E_0 = 30\,\mathrm{kV/cm}$ and $r_0 = 5\,\mu\mathrm{m}$.

Here, a spherical coordinate system is used in the calculation. Then, the radial electric field is given by

$$\overrightarrow{E_r} = E_0(\frac{r}{r_0})^3 \frac{3\cos\theta}{(1+\cos^2\theta)^{\frac{5}{2}}} \overrightarrow{e_r} \quad . \tag{2.6}$$

In order to simplify the calculation, it is assumed that the electrons will move radially. Then, the perpendicular component of the kinetic energy is given by

$$\frac{1}{2} mc^2 \beta^2_{\mathrm{ini},\perp} = \int_{r_0}^{r} e\overrightarrow{E_r} \cdot \overrightarrow{\mathrm{d}l} \quad . \tag{2.7}$$

Here, l is the distance the electron moves parallel to the cathode surface. Combining equations 2.6 and 2.7, the perpendicular velocity component $\beta_{\mathrm{ini},\perp}$ is

$$\begin{aligned} \beta_{\mathrm{ini},\perp} &= \sqrt{3}(\frac{e}{mc^2})^{\frac{1}{2}} (E_0 r_0)^{\frac{1}{2}} \sin\theta \cos^{\frac{1}{2}}\theta (1-(\frac{r_0}{r})^2)^{\frac{1}{2}} \\ &\approx \sqrt{3}(\frac{e}{mc^2})^{\frac{1}{2}} (E_0 r_0)^{\frac{1}{2}} \sin\theta \cos^{\frac{1}{2}}\theta \quad . \end{aligned} \tag{2.8}$$

Although, there is a difference between the analytical value calculated by equation 2.8 and the numerical value due to the approximation of the analytical method (as is shown in Figure 2.5(a)); the numerical calculation also confirms the relation

$$\beta_{\mathrm{ini},\perp} \propto (E_0 r_0)^{\frac{1}{2}} \quad . \tag{2.9}$$

Equation 2.9 was found to be valid for all kinds of microstructures investigated in the operating range of thermionic emission. This equation shows that an increase of the applied electric field and/or the microstructure size on the emitter surface also increases the influence of the surface roughness.

Similar to [Tsi72] and [BH97], the magnetic field is neglected due to the fact that the evaluation happens in a very small region (several tens of microns). Hence, the $\vec{E} \times \vec{B}$ drift effect on the electron at the emitter surface is not significant in such small region.

In addition, the Schottky effect, as described in equation 1.3, can also influence the distribution of $\beta_{\text{ini},\perp}$. When the Schottky effect is taken into account, the current density becomes a function of the electric field and hence of the position of the emission. It should be noted that the studied current density is in the temperature limited regime. As is shown in Figure 2.6, the Schottky effect has a limited influence. Therefore, it is also neglected in the model.

The data plotted in Figure 2.5 and equation 2.9 are stored in the code ESRAY. The stored database is used to define the transverse velocity of the electrons emitted from the emitter ring using the Monte Carlo method [Has70]. The Monte Carlo method uses a large number of repeated random samplings to obtain numerical results. Using the Monte Carlo method, the code randomly chooses for each emitted electron the kind of microstructure and a relative position in the microstructure geometry. Then, considering the electric field calculated by the code and the microstructure size, which is defined by the user, the code scales the initial $\beta_{\text{ini},\perp}$ using the scaling given by equation 2.9. Typical trajectories of the electrons close to the emitter surface in the presence of surface roughness are shown in Figure 2.7.

The microstructure size r_0 used in Figure 2.7 is $5\,\mu\text{m}$. All six kinds of microstructures evenly distributed with the same size of r_0 used in this calculation. It is shown that the surface roughness will disturb and mix the electron trajectories in the cathode region.

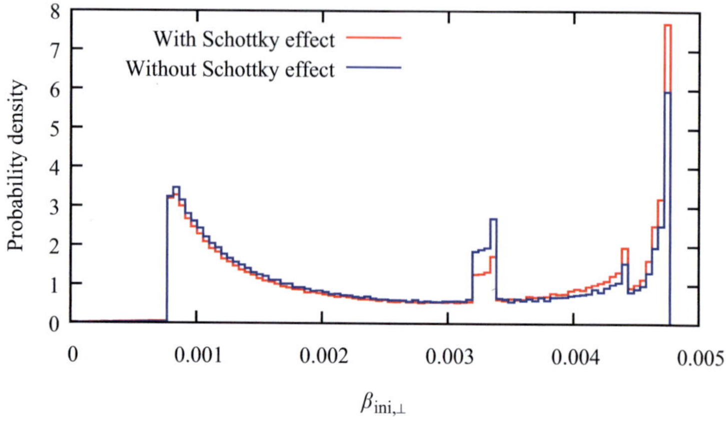

(a) Hemispherical microstructure

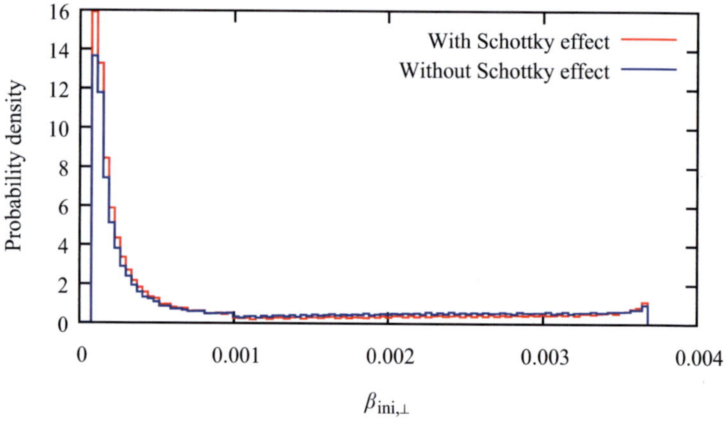

(b) Inverse hemispherical microstructure

Figure 2.6: Probability density of $\beta_{\mathrm{ini},\perp}$ from hemispherical and inverse hemispherical microstructures for $E_0 = 30$ kV/cm and $r_0 = 5$ µm. The length of the planar area is also 5 µm.

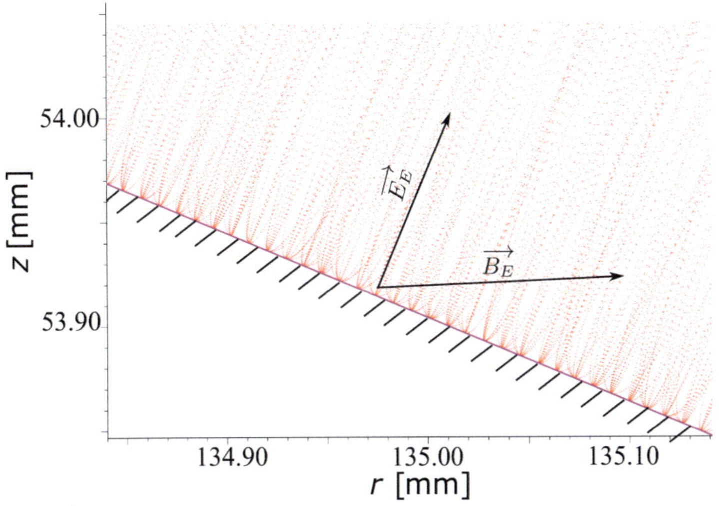

Figure 2.7: Electron trajectories at the emitter surface in the presence of surface roughness.

2.3 Influence of surface roughness on the gyrotron operation

The influence of the surface roughness on electron beam quality, gyrotron efficiency, and mode competition is studied separately. Two MIG designs for the EU 1 MW, 170 GHz ITER gyrotron [PIS+08] have been investigated. MIG-1 is the initial design of the MIG, while MIG-2 is the optimization of MIG-1, which can suppress the generation of Electron Beam Halo (EBH, Chapter 6) in the gyrotron MIG. During the modification process, the profile of the cathode surface and the configuration of the magnet coils has been changed. Therefore, the electric field and magnetic field distribution is different for these two MIGs. The electrostatic code ESRAY is used to calculate the influence of the emitter surface roughness on the electron beam quality, while the gyrotron efficiency and mode competition is studied using the cavity interaction code EURIDICE.

2.3.1 Influences of surface roughness on beam quality

The influence of the size of the microstructures on the pitch factor of the electron beam at the gyrotron cavity is shown in Figure 2.8. Both short-pulse (non-neutralized beam) and long-pulse (fully neutralized beam) operation cases for both guns have been studied. During long-pulse operation, the electron beam will be neutralized by the ions that come from the ionization of the residual gas in the MIG [PHA$^+$08]. The neutralization moderates the space charge influence on the beam tunnel and cavity region and the average α of the electron beam decreases.

In particular, the average pitch factor α is shown in Figure 2.8(a), while the pitch factor distribution of MIG-1 under long-pulse operation is presented in Figure 2.8(b). The average pitch factors α of these two guns are set to have the same value in the neutralized case. The pitch factor α increases with the size of microstructures. For the short-pulse cases, a fraction of the electrons will be reflected if the microstructure size is larger than 17 μm. The critical size of the microstructures concerning the generation of back reflected electrons is increased to 27 μm in the long-pulse case. The slightly different behaviour between these two guns results from the different magnetic and electric field configurations. The α spread, especially the maximum α value, significantly increases with the microstructure size.

Figure 2.9 shows the relation between the microstructure size and the RMS value of the transverse velocity spread $\delta\beta_\perp$ for the short-pulse and long-pulse cases, for both guns, also compared with the theoretical results based on equation 2.3.

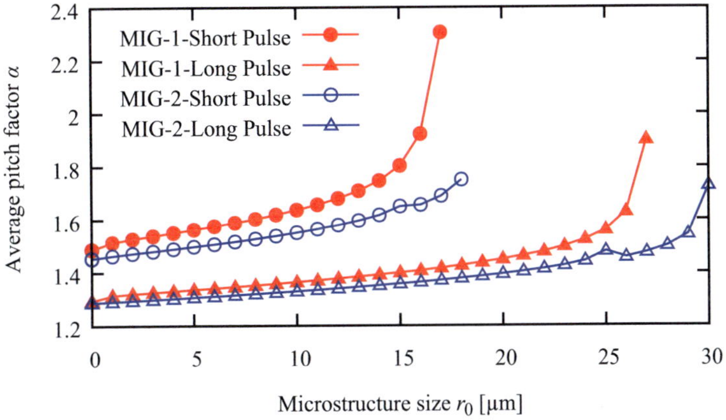

(a) Influence of surface roughness on average α

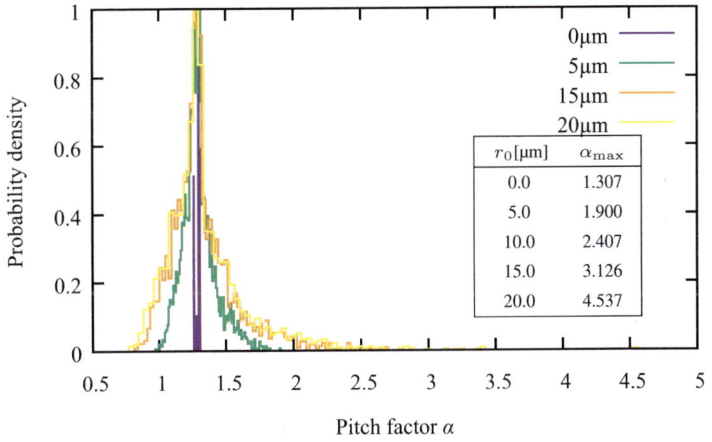

(b) Influence of surface roughness on α distribution

Figure 2.8: Influence of surface roughness on α of two EU 1 MW, 170 GHz MIG designs.

A factor of 2.6 dividing the results of equation 2.3 was considered in the theoretical result due to the different definition of the velocity spread[1]. All four curves in Figure 2.9 indicate that $\delta\beta_\perp$ is proportional to $r_0^{1/2}$ which is in agreement with the theoretical result of equation 2.3.

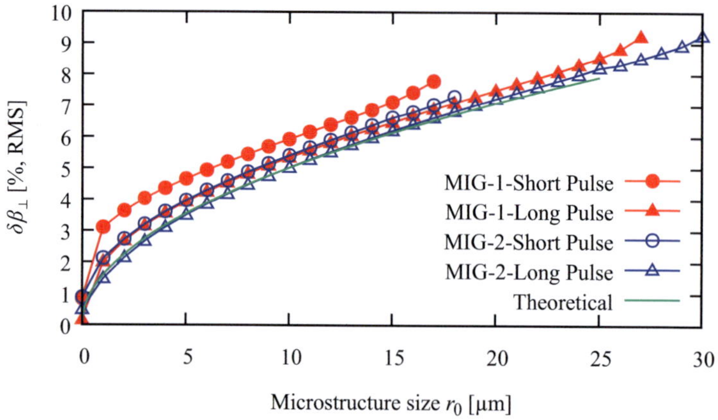

Figure 2.9: Influence of surface roughness on transverse velocity spread $\delta\beta_\perp$.

Figure 2.10 shows the relation between the microstructure size and the energy spread in the gyrotron cavity. The calculation for the short-pulse operation is stopped at 17 μm, as trapped electrons are generated for the larger size of the surface roughness. For both the neutralized case and non-neutralized case, the surface roughness has a limited influence on the energy spread. For the non-neutralized case, surface roughness can even decrease

[1] In equation 2.3 the transverse velocity spread is defined as the difference between 90 % and 10 % quantiles, which is normalized by the mean value: $\delta\beta_\perp = \frac{F(0.9)-F(0.1)}{\mu}$. While the RMS spread is define as $\delta\beta_\perp = \frac{\sigma}{\mu}$. Assuming the velocity spread has a Gaussian distribution gives $\frac{F(0.9)-F(0.1)}{\sigma} \approx 2.6$.

the energy spread. This is because the surface roughness will increase the phase mixing of electrons in the beam.

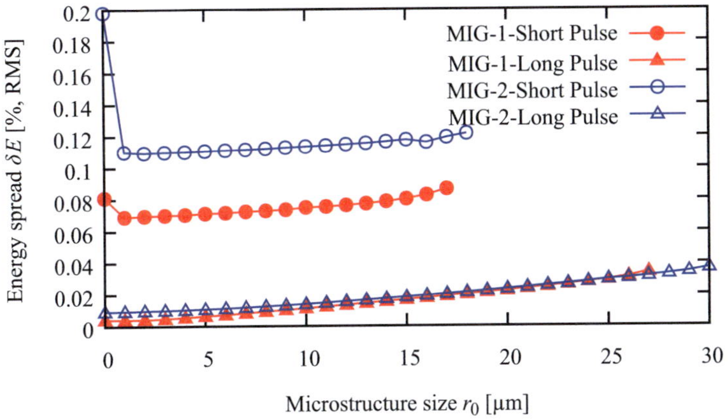

Figure 2.10: Influence of surface roughness on energy spread.

2.3.2 Influence of surface roughness on gyrotron efficiency

The influence of surface roughness on gyrotron efficiency during long-pulse operation (neutralized case) is calculated using the cavity interaction code EURIDICE. The electron beam properties at the entrance of the cavity are calculated by ESRAY. Single-mode calculation results for the electronic efficiency η_{elec} and the total efficiency η_{tot} (with single-stage depressed collector) are shown in Figure 2.11.

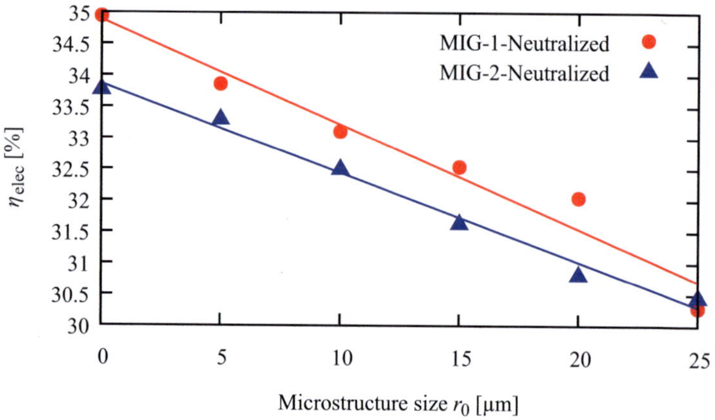

(a) Without depressed collector

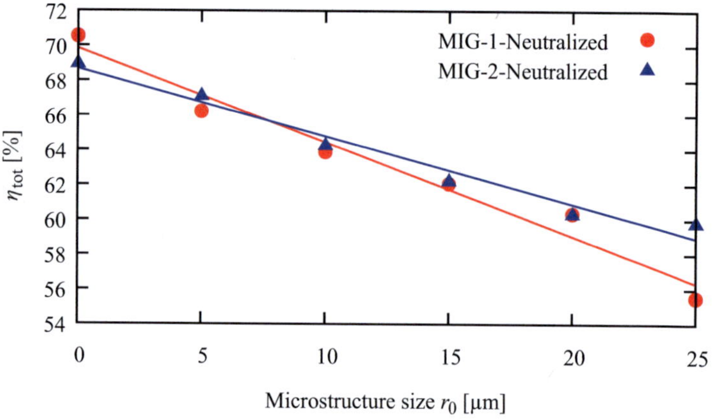

(b) With depressed collector

Figure 2.11: Influence of surface roughness on gyrotron efficiency.

In particular, the relation between the microstructure size and the beam-wave interaction efficiency is shown in Figure 2.11(a). An increase of the emitter surface microstructure size from 0 µm to 25 µm results in a decrease of the gyrotron interaction efficiency from 34.9 % to 30.3 % for MIG-1 and from 33.6 % to 30.5 % for MIG-2. The efficiency is approximately linearly dependent on the microstructure size. This is in agreement with equation 2.4. Then, the relation between the microstructure size and the interaction efficiency can be simplified as

$$\eta_{\mathrm{elec}} = \eta_{\mathrm{elec,max}}(1 - K_r r_0) \quad . \tag{2.10}$$

Here, K_r is the coefficient of the surface roughness. For these two guns, 1 µm of microstructure size increase corresponds to 0.15% decrease of the interaction efficiency.

Figure 2.11(b) shows the relation between the microstructure size and the maximum achievable total efficiency in the case of single-stage depressed collector operation. The relation between the interaction efficiency and the total efficiency is given in equation 1.23. The maximum V_{col}, resulting in the maximum η_{tot}, is calculated assuming that the acceptable portion of electrons reflected by the depressed collector is below 1 %. The total efficiency η_{tot} is significantly decreased by 15 % and 9 %, respectively, as the microstructure size increases from 0 µm to 25 µm. Calculations also show that the increase of the microstructure size will lead to an approximately 5.5 kV decrease in the maximum allowed depressed collector voltage at 25 µm.

2.3.3 Influence of surface roughness on gyrotron start-up

The influence of the surface roughness on the gyrotron start-up was also studied. The electron beam parameters calculated by the code ESRAY during the start-up of MIG-2 are shown in Figure 2.12. The accelerating voltage is increased from 47 kV to 85.1 kV in 8 µs in the simulation. At each time step, the electrons are imported from ESRAY to EURIDICE for the calculation of beam-wave interaction.

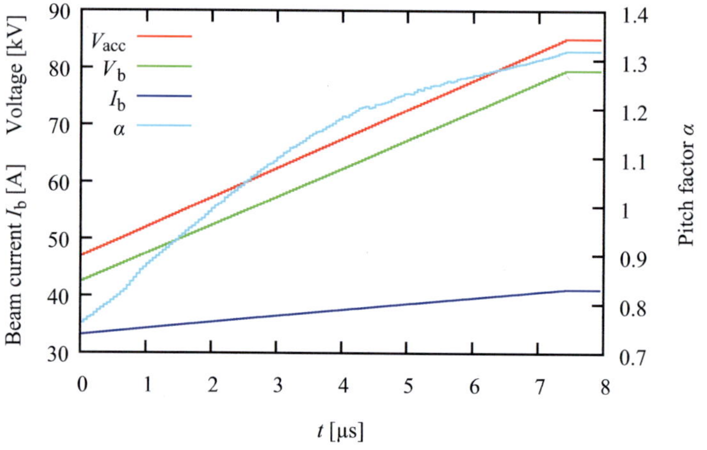

Figure 2.12: Start up process of MIG-2. V_{acc}: accelerating voltage, V_b: beam voltage, I_b: beam current, α: pitch factor.

2000 electrons are used for the multimode calculation in EURIDICE. The results are shown in Figure 2.13. Figure 2.13(a) shows the output power during the whole simulation period while Figure 2.13(b) gives the zoom of the simulation result between 5 µs and 7 µs.

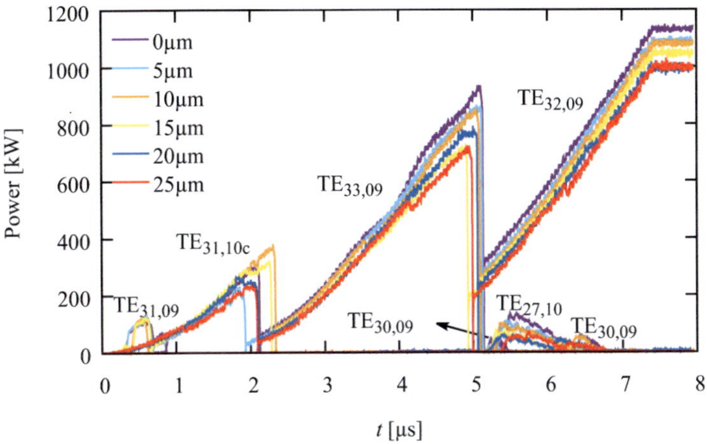

(a) Mode competition between the whole simulation period

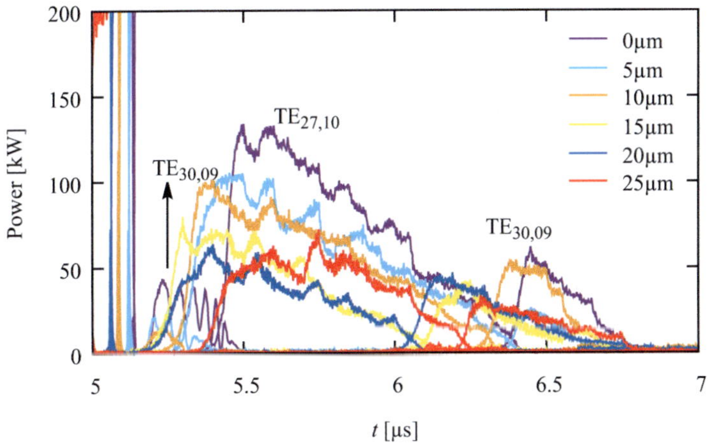

(b) Mode competition between 5 μs and 7 μs

Figure 2.13: Influence of surface roughness on the start-up of MIG-2.

The $TE_{32,09}$ mode is the nominal cavity mode. To fully consider the mode competition effect, 33 additional competing modes are used in the calculations. The final output power decreases with the increase of microstructure size, but the nominal mode is excited at a lower voltage due to the increase of the microstructure size.

Between 5 µs and 7 µs, the $TE_{30,09}$ mode can start two times when the microstructure size is smaller than 10 µm, but when the microstructure size is larger than 10 µm, the first excitation of the $TE_{30,09}$ mode disappears. In addition, the excitation of the mode $TE_{31,09}$ at the beginning of the voltage ramp disappears for microstructure size above 20 µm. This phenomenon indicates that weak competing modes are more sensitive to the beam quality, which confirms that velocity spread can suppress weak competing modes and facilitate mode selection [ZZM06].

2.4 Summary

The influence of emitter surface roughness on gyrotron operation was investigated using a model which was developed for the codes ESRAY and ESPIC. The surface roughness model uses an initial velocity distribution to simulate the influence of surface roughness. With this model, one can calculate its influence on gyrotrons without modelling the real surface roughness and increase the mesh number in the code. This feature can decrease the calculation time significantly. Magnetron injection gun simulations show that the increase of the velocity spread is proportional to $(E_0 r_0)^{\frac{1}{2}}$, where r_0 represents the microstructure size and E_0 is the average applied electric field strength on the emitter surface. A typical surface roughness value $r_0 = 5\,\mu\mathrm{m}$ can result in an increase in transverse velocity spread $\delta\beta_\perp$ by a fraction of approximately $4\,\%$ (from $0.15\,\%$ to $4\,\%$) and a decrease in the interaction efficiency η_{elec} by a fraction of approximately $1\,\%$ (from $34.9\,\%$ to $33.8\,\%$). The influence of initial thermal energy on the velocity spread is much smaller than the influence of surface roughness.

A very rough emitter surface could result in the generation of harmful back reflected electrons. Single-mode cavity calculation results indicate that the decrease of microwave efficiency is proportional to the microstructure size r_0. Magnetron injection gun and cavity simulation results are in agreement with the conclusion of the previous theoretical analysis. The gyrotron start-up calculation shows that although the increase of microstructure size will decrease the output power it can also suppress some weak competing modes. The surface roughness model derived for the codes ESRAY and ESPIC in this chapter is an important tool to assist in the advanced design of more robust magnetron injection guns for gyrotrons in the future. The surface roughness model is used in the investigation on two harmful phenomena in gyrotron: Low Frequency Oscillations (LFOs, Chapter 5) and Electron Beam Halo (EBH, Chapter 6).

3 Evaluation of emission inhomogeneity from current voltage characteristics

Besides surface roughness, current emission inhomogeneity of emitters can also result in a degradation of the electron beam quality. In this Chapter, an introduction to the emission inhomogeneity is given first. After a review of three existing evaluating methods of emission inhomogeneity, a novel definition of emission inhomogeneity and evaluating method is described. Then, emission inhomogeneities of three different types of gyrotron oscillators at different emitter temperatures are tested. Subsequently, the influence of emission inhomogeneity on gyrotron efficiency is numerically calculated using the 3D code ARIADNE [PV04] and EURIDICE. Finally, the upper acceptable limits for emitter inhomogeneity of the work function, temperature, and electric field are given at the end of this chapter.

3.1 Introduction

Emission inhomogeneity of the annular electron beam does have a significant effect on the gyrotron operation. The emission inhomogeneity is a comprehensive result of temperature, work function and electric field distribution on the emitter surface. It has been experimentally tested that the emission inhomogeneity decreases the gyrotron efficiency [GKV$^+$97] and causes Low Frequency Oscillations (LFOs) in the gyrotron [LPS$^+$06b].

Numerical simulation also shows that with a typical inhomogeneity of emission the gyrotron efficiency is 2 %-3 % lower than that for ideally uniform emission [NVB+01]. There are several kinds of experimental methods to measure the emission inhomogeneity: Scanning Auger Microprobe (SAM) [EK81], Rotating Current Probe (RCP) [ATS05] and Miram curve [MIR+04]. SAM measures emitter work-function distribution based on patches of different work function giving rise to different onsets of secondary electron emission. RCP can measure the gyrotron angular current distribution by rotating the current probe. The Miram curve translates measured temperature-current data into work-function-distribution-like curves. However, all of these methods are either not suitable for MIGs (it is difficult to measure the emitter temperature in MIGs) or need additional measurement devices. To evaluate the emission inhomogeneity from the Current Voltage Characteristics (CVC) of the MIG is the easiest and fastest method for high power gyrotrons. As introduced in Section 1.4.1, thermionic emitters are used in high power gyrotrons. With the increase of the electric field on the emitter surface, a thermionic emitter will first work under the space charge limited region and then works in the saturation region or temperature limited region, as is shown in the blue line of Figure 3.1.

For ideal uniform emitters, these two regions are connected at a single point. The real situation is that properties of the emitter are different from one part to another, and the current density obtained at different spots of the emitter varies. As a result, the space charge and the temperature limited regions are connected with a transition region instead of being connected at a single point, as is shown in the red line of Figure 3.1. Therefore, the information on emission inhomogeneity is included in the CVC, especially in the characteristics of the transition region. Evaluating the emission inhomogeneity from the CVC does not need any additional measurement devices besides the usual current voltage measurement device. Therefore, it is quite convenient to measure the emission inhomogeneity during the experiment. This method can be used to study gyrotron cathode emitter aging during the gyrotron lifetime.

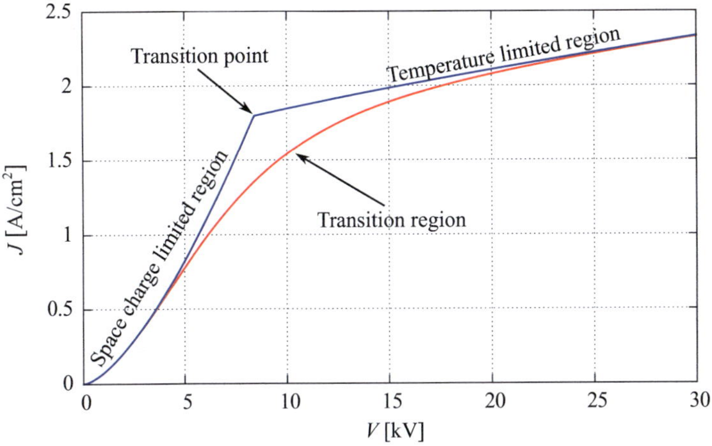

Figure 3.1: Current voltage characteristic of a planar diode.

3.2 Modification of emission inhomogeneity models

Generally, there are two methods that were initially developed for a parallel plate diode (Figure 3.2(a)), which can be used to investigate the emission inhomogeneity using the CVC. According to their different assumptions, these two models are named Khodnevich model (saturation current density distribution model) [Kho69, GKV+97, GGK+99] and Tonnerre model (work function distribution model) [TBPS83], respectively. However, the gyrotron MIG emitter has an annular geometry, as is shown in Figure 3.2(b). Due to the special geometry of the gyrotron MIG emitter, these two models can not be applied to gyrotron directly. In order to employ these two models, the Child-Langmuir equation and Richardson-Dushman equation have to be modified.

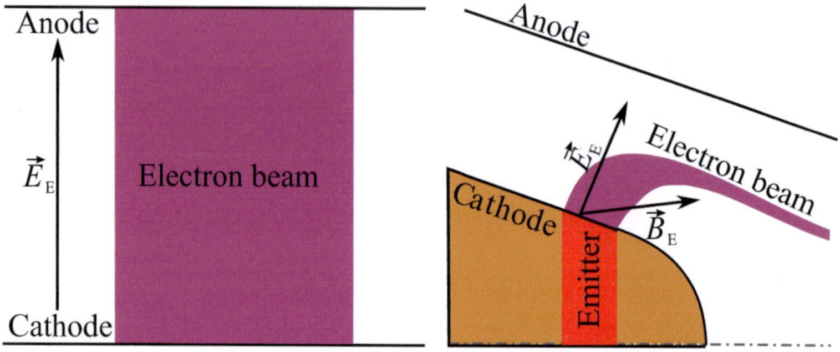

(a) Parallel plate diode

(b) Gyrotron emitter structure

Figure 3.2: Geometry of (a) parallel plate diode and (b) gyrotron MIG emitter.

In gyrotrons, the size of the emitter is limited and due to the influence of the magnetic field above the emitter, the current density is not uniformly distributed. Due to this, it is difficult to find an analytical solution for the space charge limited current. In the present work, a correction factor $\beta_{\text{C-L}}$ is used to modify the Child-Langmuir equation 1.2. Then the modified Child-Langmuir equation is given by

$$J_{\text{C-L}} = \beta_{\text{C-L}} A_{\text{C-L}} V_0^{\frac{3}{2}} \quad . \tag{3.1}$$

For the planar diode, this correction factor is one. For gyrotrons, there are two effects that influence the correction factor. Firstly, electrons are emitted only from the small ring-shaped emitter surface in gyrotron, while in the planar diode, electrons are emitted from the whole cathode surface. Therefore, there is no space charge effect from other parts of the cathode, which limits the electrons current from the emitter ring. The correction factor will be larger than one due to this effect. Secondly, the magnetic field will change the direction of electrons during their movement towards the anode. The thickness of the electron beam is reduced at the corner of the beam trajectory. The reduced beam thickness results in a larger space charge density in this region. The correction factor will be decreased due to this effect.

The electric field on the emitter surface of a planar diode is $E_0 = V_0/d$. Due to the special geometry of the emitter surface, the electric field E is always larger than E_0. The relation between E and E_0 is given by

$$E = \beta_E E_0 \tag{3.2}$$

where β_E is the field enhancement factor. The modified Richardson-Dushman equation with Schottky correction is given by

$$\begin{aligned} J_\mathrm{S} &= A_\mathrm{g} T^2 e^{-\frac{W - \triangle W}{kT}} \\ \triangle W &= \sqrt{\frac{\beta_E e^3 E_0}{4\pi\varepsilon_0}} \quad . \end{aligned} \tag{3.3}$$

3.2.1 Khodnevich model

The Khodnevich model [Kho69] does not consider the Schottky effect in the temperature limited region. Equation 1.3 reduces to the Richardson-Dushman equation

$$J_\mathrm{R} = A_\mathrm{g} T^2 e^{-\frac{W}{kT}} \tag{3.4}$$

The CVCs of a uniform emitter with and without consideration of the Schottky effect are shown in Figure 3.3.

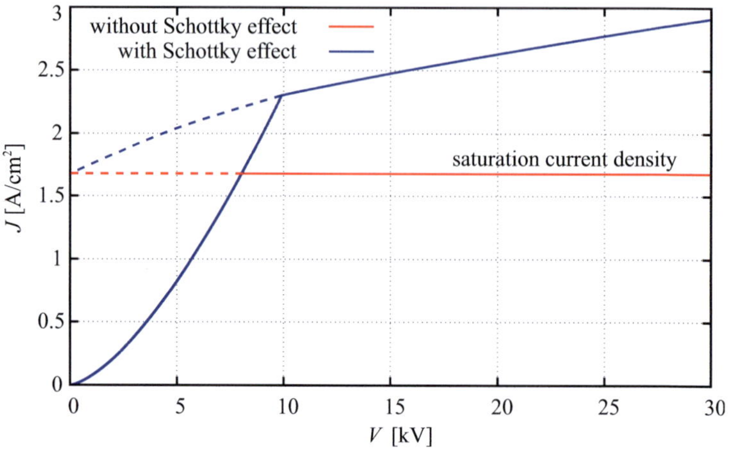

Figure 3.3: Comparison of the current voltage characteristics with (blue line) and without (red line) the Schottky effect.

In the Khodnevich model, the emitter has a distribution of the spatial saturation current density, which can be described by equation 3.4. The reasons that cause the inhomogeneity of the current density are the distribution of work function W and temperature T around the emitter surface. With a given voltage, every part of the emitter has the same space charge limited current density, but different saturation current density. These parts of the emitter, which have a higher saturation current density than the space charge limited current density work under the space charge limited region since the total current will be limited by the space charge effect. The other parts, which have lower saturation current densities than the space charge limited current density, will work under the temperature limited region.

The emitter surface area S and the total current I can be separated into two parts

$$S = S_{\text{C-L}}(V) + S_{\text{R}}(V) \tag{3.5}$$

$$I(V) = I_{\text{C-L}}(V) + I_{\text{R}}(V) \tag{3.6}$$

where $S_{\text{C-L}}$ and $I_{\text{C-L}}$ are the area and current that work under the space charge limited region while S_{R} and I_{R} are the area and current that work under the temperature limited region. When the applied voltage $V \to 0$, the whole emitter works under space charge limited region and $S_{\text{R}} \to 0$. When the applied voltage $V \to \infty$, the whole emitter works under temperature limited region and $S_{\text{R}} = S$.

The first part of the total current (space charge limited current) is

$$\begin{aligned} I_{\text{C-L}}(V) &= \int_0^{S_{\text{C-L}}(V)} J_{\text{C-L}}(V)dS \\ &= \int_0^{S_{\text{C-L}}(V)} \beta_{\text{C-L}} A_{\text{C-L}} V^{\frac{3}{2}} dS \\ &= \beta_{\text{C-L}} A_{\text{C-L}} V^{\frac{3}{2}} S_{\text{C-L}}(V) \quad . \end{aligned} \tag{3.7}$$

In order to simplify the calculation, the substitution $u := V^{\frac{3}{2}}$ is used. Then the space charge limited current is

$$I_{\text{C-L}}(u) = \beta_{\text{C-L}} A_{\text{C-L}} u S_{\text{C-L}}(u) \quad . \tag{3.8}$$

The second part of the total current (temperature limited current) is

$$\begin{aligned} I_{\text{R}}(V) &= \int_0^{S_{\text{R}}(V)} J_{\text{R}}(V)dS \\ &= \int_0^{S_{\text{R}}(u)} J_{\text{R}}(u)dS \quad . \end{aligned} \tag{3.9}$$

The distribution of $S_{\mathrm{R}}(u)$ is defined by

$$f(u) := \frac{\mathrm{d}S_{\mathrm{R}}(u)}{du} \quad . \tag{3.10}$$

With the above definition of $f(u)$, the area of the emitter surface that works under space charge limited region under u can be derived by

$$S_{\mathrm{R}}(u) = \int_0^u f(u^{'})du^{'} \quad . \tag{3.11}$$

Combining equation 3.9 and equation 3.10 leads to

$$I_{\mathrm{R}}(u) = \int_0^u J_{\mathrm{R}}(u^{'})f(u^{'})du^{'} \quad . \tag{3.12}$$

With the expressions of equation 3.7 and equation 3.12, equation 3.6 can be expanded to

$$I(u) = \beta_{\text{C-L}}A_{\text{C-L}}uS_{\text{C-L}}(u) + \int_0^u J_{\mathrm{R}}(u^{'})f(u^{'})du^{'} \quad . \tag{3.13}$$

Combining equation 3.24 with equation 3.13, one gets

$$\begin{aligned} I(u) &= \beta_{\text{C-L}}A_{\text{C-L}}u(S - S_{\mathrm{R}}(u)) + \int_0^u J_{\mathrm{R}}(u^{'})f(u^{'})du^{'} \\ &= \beta_{\text{C-L}}A_{\text{C-L}}uS - \beta_{\text{C-L}}A_{\text{C-L}}u\int_0^u f(u^{'})du^{'} + \int_0^u J_{\mathrm{R}}(u^{'})f(u^{'})du^{'} \end{aligned} \tag{3.14}$$

Because the Schottky effect is neglected in this model, the current density will always be equal to the transition current density, which means

$$J_{\mathrm{R}}(u) = \beta_{\text{C-L}}A_{\text{C-L}}u \quad . \tag{3.15}$$

Then the total current will be

$$
\begin{aligned}
I(u) &= \beta_{\text{C-L}} A_{\text{C-L}} u S - \beta_{\text{C-L}} A_{\text{C-L}} u \int_0^u f(u^{'}) du^{'} + \beta_{\text{C-L}} A_{\text{C-L}} \int_0^u u^{'} f(u^{'}) du^{'} \\
&= \beta_{\text{C-L}} A_{\text{C-L}} u S - \beta_{\text{C-L}} A_{\text{C-L}} \int_0^u \int_0^{u'} f(u^{''}) du^{''} du' \quad . \qquad (3.16)
\end{aligned}
$$

Using the definition of equation

$$
S_{\text{R}}(u) = \int_0^u f(u^{'}) du^{'} = S_{\text{R}}(J_{\text{R}}) = \int_0^{J_R} F(j_{\text{R}}) dj_{\text{R}} \qquad (3.17)
$$

where $F(j_{\text{R}})$ is the distribution of the transition current density which is also named effective emission inhomogeneity.

Then the total current becomes

$$
I(u) = \beta_{\text{C-L}} A_{\text{C-L}} u S - \beta_{\text{C-L}} A_{\text{C-L}} \int_0^u \int_0^{J_R} F(j_{\text{R}}) dj_{\text{R}} du^{'} \qquad (3.18)
$$

and one can get the first and second derivative of the total current from equation 3.18

$$
\begin{aligned}
\frac{\mathrm{d}I}{\mathrm{d}u} &= \beta_{\text{C-L}} A_{\text{C-L}} S - \beta_{\text{C-L}} A_{\text{C-L}} \int_0^{J_R} F(j_{\text{R}}) dj_{\text{R}} \qquad (3.19) \\
\frac{\mathrm{d}^2 I}{\mathrm{d}u^2} &= -(\beta_{\text{C-L}} A_{\text{C-L}})^2 F(j_{\text{R}}) \qquad (3.20)
\end{aligned}
$$

From equation 3.20 one gets

$$
F(j_{\text{R}}) = -\frac{1}{(\beta_{\text{C-L}} A_{\text{C-L}})^2} \frac{\mathrm{d}^2 I}{\mathrm{d}u^2} \qquad (3.21)
$$

From the measured current voltage curve, one can calculate the distribution of the transition current density according to equation 3.21.

Furthermore, Combination of equations 3.17 and 3.19 results in

$$\begin{aligned}\frac{\mathrm{d}I}{\mathrm{d}u} &= \beta_{\text{C-L}} A_{\text{C-L}} S - \beta_{\text{C-L}} A_{\text{C-L}} \int_0^{J_R} F(j_{\text{R}}) dj_{\text{R}} \\ &= \beta_{\text{C-L}} A_{\text{C-L}} S - \beta_{\text{C-L}} A_{\text{C-L}} S_{\text{R}} \\ &= \beta_{\text{C-L}} A_{\text{C-L}} S_{\text{C-L}} \quad . \end{aligned} \tag{3.22}$$

Then the areas which work under space charge limited region and temperature limited region are

$$S_{\text{C-L}} = \frac{1}{\beta_{\text{C-L}} A_{\text{C-L}}} \frac{\mathrm{d}I}{\mathrm{d}u} \tag{3.23}$$

$$S_{\text{R}} = S - \frac{1}{\beta_{\text{C-L}} A_{\text{C-L}}} \frac{\mathrm{d}I}{\mathrm{d}u} \quad . \tag{3.24}$$

Figure 3.4 shows the calculation result for a parallel plate diode. The CVC curve (a) is generated using a random number generator assuming that the work function has a normal distribution, the average work function W_0 of the emitter is 2.2 eV and σ_W is 0.075 eV. The calculation process of the saturation current density distribution is: after the CVC curve is measured (a), the first (b) and second (c) derivative is made. From the $\mathrm{d}^2 I/\mathrm{d}u^2$ curve shown in Figure 3.4, one can easily find the final saturation current density distribution using equation 3.15.

To compare the quality of different emitters, one can normalize current and voltage to their values I_m and u_m corresponding to the maximum of the density distribution function [GGK+99]. Those parts of the cathode surface working in the space charge regime and in the saturation regime are normalized to the whole surface area of the cathode. Then the system of variables will be

$$\hat{u} = \frac{u}{u_m}, \quad \hat{I} = \frac{I}{I_m}, \quad \hat{S}_{\text{C-L}} = \frac{S_{\text{C-L}}}{S}, \quad \hat{S}_{\text{R}} = \frac{S_{\text{R}}}{S} \quad . \tag{3.25}$$

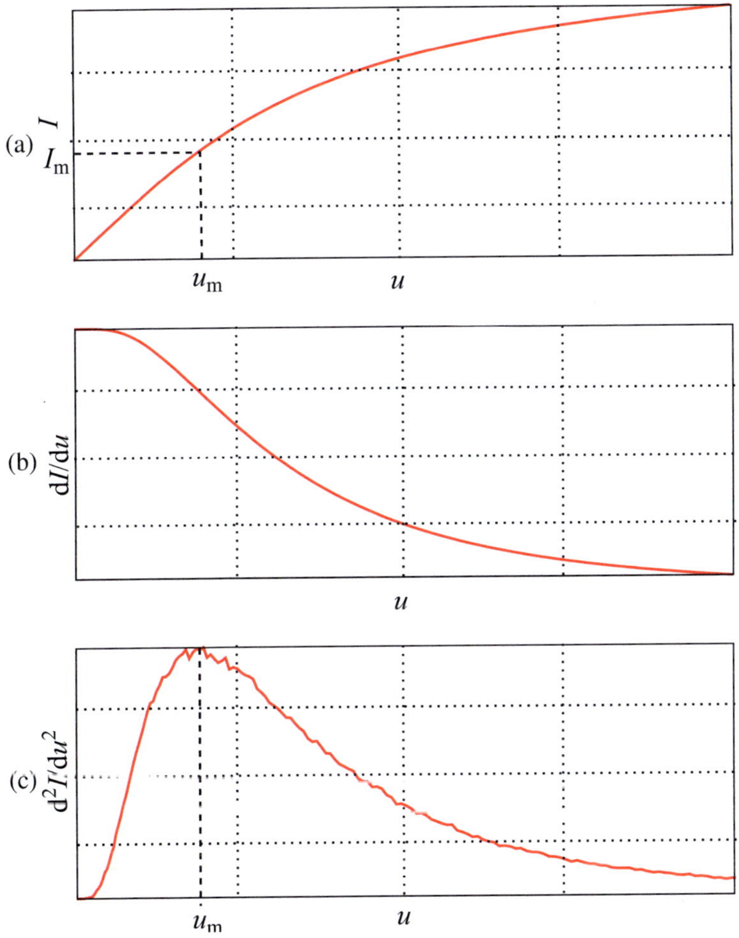

Figure 3.4: Calculation process of the saturation current density distribution model. (a): Measured/Calculated CVC curve, (b): first derivative, (c): second derivative.

The normalized distribution of the effective emission inhomogeneity will have the following property

$$\int_0^\infty F(\hat{j}_\mathrm{R})d\hat{j}_\mathrm{R} = 1 \quad . \tag{3.26}$$

Usually, the normalized distribution density function is fitted by the normal distribution law, which is

$$F(\hat{j}_\mathrm{R}) = \frac{1}{\sqrt{2\pi}\sigma_{\hat{j}_\mathrm{R}}} exp(-\frac{(\hat{j}_\mathrm{R}-1)^2}{2\sigma^2_{\hat{j}_\mathrm{R}}}) \quad . \tag{3.27}$$

The fitted result of $\sigma_{\hat{j}_\mathrm{R}}$ is called the normalized effective emission inhomogeneity.

3.2.2 Tonnerre model

In the Tonnerre model [TBPS83], it is assumed that the smooth transition between the space charge limited region and the temperature limited region of the CVC curve results from the work function distribution of the emitter. The emitter surface is separated into elementary parts with fractional area θ_i and work function W_i. At a given voltage V, parts of the cathode surface that have a work function W_i higher than a particular value W_t work in the temperature limited mode, whereas those parts having a work function W_i lower than W_t still work in the space charge mode. For these parts, which work under the transition point, their space charge limited current density is equal to the temperature limited current density, which gives

$$W_\mathrm{t} = \sqrt{\frac{\beta_E e^3 V}{4\pi d\varepsilon_0}} - kT\ln(\frac{\beta_\text{C-L} A_\text{C-L} V^{\frac{3}{2}}}{A_\mathrm{g} T^2}) \quad . \tag{3.28}$$

The average current density J_V is the mean of the current densities emitted in both modes by the total surface. To simplify the summation, it is assumed that each part of the emitter has the same A_{C-L} and A_g value, neglecting the influence of neighbouring elements of different areas and work functions. Then the average current density at voltage V follows

$$J_V = \sum_{W_{\mathrm{i}}=W_{\mathrm{min}}}^{W_{\mathrm{i}}=W_{\mathrm{t}}} J_{\mathrm{C\text{-}L}}\theta_{\mathrm{i}} + \sum_{W_{\mathrm{i}}=W_{\mathrm{t}}}^{W_{\mathrm{i}}=W_{\mathrm{max}}} J_{\mathrm{Si}}\theta_{\mathrm{i}} \quad . \tag{3.29}$$

θ_{i} is normalized, which gives

$$\sum_{W_{\mathrm{i}}=W_{\mathrm{min}}}^{W_{\mathrm{i}}=W_{\mathrm{max}}} \theta_{\mathrm{i}} = 1 \tag{3.30}$$

where $J_{\mathrm{C\text{-}L}}$ and J_{S} are defined by equations 1.2 and 1.3, respectively.

To simplify the calculation one can define

$$\alpha_T := \frac{\sqrt{\frac{\beta_E e^3}{4d\pi\varepsilon_0}}}{kT} \quad . \tag{3.31}$$

Then, the temperature limited current density for each part of the emitter is

$$J_{\mathrm{Si}} = A_{\mathrm{g}} T^2 e^{\alpha_T V^{\frac{1}{2}}} e^{-\frac{W_{\mathrm{i}}}{kT}} \quad , \tag{3.32}$$

and the average current density J_{V} is given by the equation

$$J_{\mathrm{V}} = \beta_{\mathrm{C\text{-}L}} A_{\mathrm{C\text{-}L}} V^{\frac{3}{2}} \sum_{W_{\mathrm{i}}=W_{\mathrm{min}}}^{W_{\mathrm{i}}=W_{\mathrm{t}}} \theta_{\mathrm{i}} + A_{\mathrm{g}} T^2 e^{\alpha_T V^{\frac{1}{2}}} \sum_{W_{\mathrm{i}}=W_{\mathrm{t}}}^{W_{\mathrm{i}}=W_{\mathrm{max}}} \theta_{\mathrm{i}} e^{-\frac{W_{\mathrm{i}}}{kT}} \quad . \tag{3.33}$$

In the Tonnere model it is supposed that the work function has a distribution $D(W)$ which is normalized as

$$\int_{W_{\mathrm{min}}}^{W_{\mathrm{max}}} D(W)dW = 1 \tag{3.34}$$

where $D(W)dW = \theta_i$ is the fraction of the cathode area having a work function whose value is between W and $W + dW$.

Then equation 3.33 becomes

$$J_{\mathrm{V}} = \beta_{\text{C-L}} A_{\text{C-L}} V^{\frac{3}{2}} \int_{W_{\min}}^{W_{\mathrm{t}}} D(W)dW + A_{\mathrm{g}} T^2 e^{\alpha_T V^{\frac{1}{2}}} \int_{W_{\mathrm{t}}}^{W_{\max}} D(W) e^{-\frac{W}{kT}} dW \tag{3.35}$$

and the first derivative of J_{V} over $\mathrm{d}V^{\frac{1}{2}}$ is

$$\begin{aligned}\frac{\mathrm{d}J_{\mathrm{V}}}{\mathrm{d}V^{\frac{1}{2}}} =& 3\beta_{\text{C-L}} A_{\text{C-L}} V \int_{W_{\min}}^{W_{\mathrm{t}}} D(W)dW + \alpha_T A_{\mathrm{g}} T^2 e^{\alpha_T V^{\frac{1}{2}}} \int_{W_{\mathrm{t}}}^{W_{\max}} D(W) e^{-\frac{W}{kT}} dW \\ &+ (\beta_{\text{C-L}} A_{\text{C-L}} V^{\frac{3}{2}} - A_{\mathrm{g}} T^2 e^{\alpha_T V^{\frac{1}{2}}} e^{-\frac{W_{\mathrm{t}}}{kT}}) D(W_{\mathrm{t}}) \frac{dW_{\mathrm{t}}}{dV^{\frac{1}{2}}}\end{aligned} \tag{3.36}$$

According to the definition of W_{t} in equation 3.28, one gets

$$\beta_{\text{C-L}} A_{\text{C-L}} V^{\frac{3}{2}} - A_{\mathrm{g}} T^2 e^{\alpha_T V^{\frac{1}{2}}} e^{-\frac{W_{\mathrm{t}}}{kT}} = 0 \quad . \tag{3.37}$$

Then, equation 3.36 becomes

$$\frac{\mathrm{d}J_{\mathrm{V}}}{\mathrm{d}V^{\frac{1}{2}}} = 3\beta_{\text{C-L}} A_{\text{C-L}} V \int_{W_{\min}}^{W_{\mathrm{t}}} D(W)dW + \alpha_T A_{\mathrm{g}} T^2 e^{\alpha_T V^{\frac{1}{2}}} \int_{W_{\mathrm{t}}}^{W_{\max}} D(W) e^{-\frac{W}{kT}} dW \tag{3.38}$$

Using the definition

$$F(W) := \int_{W_{\min}}^{W_{\mathrm{t}}} D(W)dW \tag{3.39}$$

$$S(W) := \int_{W_{\mathrm{t}}}^{W_{\max}} D(W) e^{-\frac{W}{kT}} dW \quad , \tag{3.40}$$

then, equation 3.35 and equation 3.38 become

$$J_{\text{V}} = \beta_{\text{C-L}} A_{\text{C-L}} V^{\frac{3}{2}} F(W) + A_{\text{g}} T^2 e^{\alpha_{\text{T}} V^{\frac{1}{2}}} S(W) \tag{3.41}$$

$$\frac{\mathrm{d}J_{\text{V}}}{\mathrm{d}V^{\frac{1}{2}}} = 3\beta_{\text{C-L}} A_{\text{C-L}} V F(W) + \alpha_T A_{\text{g}} T^2 e^{\alpha_T V^{\frac{1}{2}}} S(W) \quad . \tag{3.42}$$

Using the definition

$$\alpha_{\text{V}} := \frac{\mathrm{d}\ln J_{\text{V}}}{\mathrm{d}V^{\frac{1}{2}}} \quad , \tag{3.43}$$

one gets

$$F(W) = \frac{\alpha_V - \alpha_T}{\frac{3}{V^{\frac{1}{2}}} - \alpha_T} \frac{J_{\text{V}}}{\beta_{\text{C-L}} A_{\text{C-L}} V^{\frac{3}{2}}} \tag{3.44}$$

$$S(W) = \frac{\frac{3}{V^{\frac{1}{2}}} - \alpha_{\text{V}}}{\frac{3}{V^{\frac{1}{2}}} - \alpha_T} \frac{J_{\text{V}}}{A_{\text{g}} T^2 e^{\alpha_T V^{\frac{1}{2}}}} \quad . \tag{3.45}$$

According to the definition of $F(W)$ in equation 3.39, one can calculate the work function distribution $D(W)$ by

$$D(W) = \frac{\mathrm{d}F(W)}{\mathrm{d}W} \quad . \tag{3.46}$$

As an example, Figure 3.5 shows the calculation process for a planar diode. The CVC curve (a) is generated using a random number generator assuming that the work function has a Gaussian distribution, the average work function W_0 of the emitter is 2.2 eV, and σ_W is 0.075 eV. The calculation process for the work function distribution is: after the measurement of the CVC the curve (a) $F(W)$ (b) is calculated according to equation 3.44. Then, $D(W)$ (c) is calculated according to equation 3.46. The discontinuity in the CVC curve will be enlarged during the derivative. The noise of $D(W)$ comes from the high requirement of second derivative on the data accuracy.

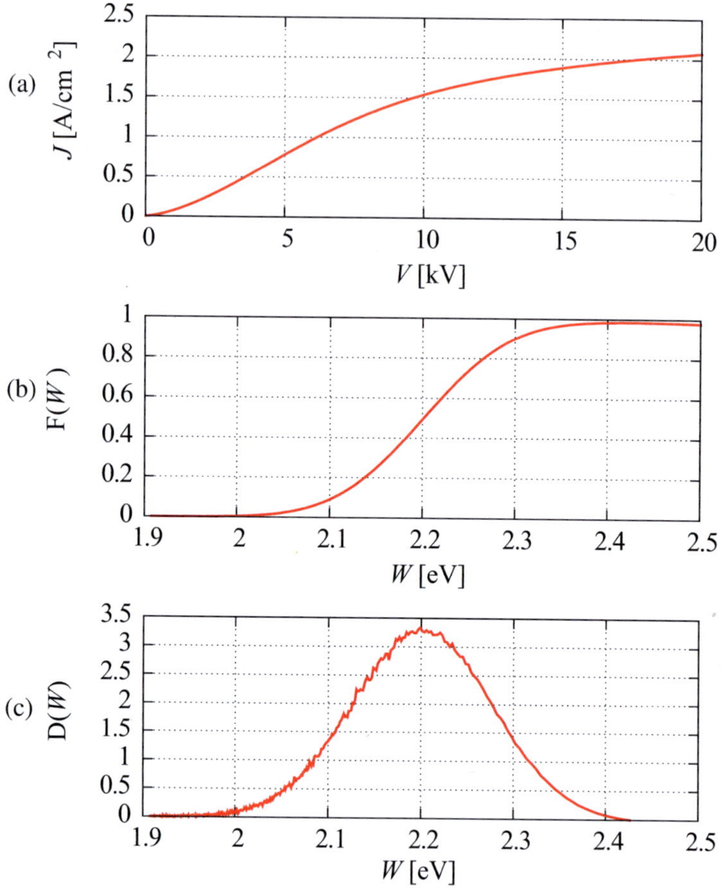

Figure 3.5: Calculation process of the work function distribution model. (a): Measured/Calcualted CVC curve, (b): $F(W)$ calculated by equation 3.44, (c): $D(W)$ calculated according to equation 3.46.

An analytical CVC curve function can be derived if the work function has the Gaussian distribution

$$D(W) = \frac{1}{\sigma_W\sqrt{2\pi}}\int_{W_{\min}}^{W} e^{-\frac{(W-W_0)^2}{2\sigma_W^2}}\,dW \quad . \tag{3.47}$$

Equation 3.35 becomes [AKT+02]

$$\begin{aligned} J_{\mathrm{V}} =& \frac{1}{2}\beta_{\text{C-L}}A_{\text{C-L}}V^{\frac{3}{2}}[1+\operatorname{erf}(\frac{W_{\mathrm{t}}-W_0}{\sigma_W\sqrt{2}})]+ \\ & \frac{1}{2}A_{\mathrm{g}}T^2[1-\operatorname{erf}(\frac{W_{\mathrm{t}}-W_0+\frac{\sigma_W^2}{kT}}{\sigma_W\sqrt{2}})]e^{\frac{1}{kT}(\sqrt{\frac{e^3V}{4\pi d\varepsilon_0}}-W_0+\frac{\sigma_W^2}{2kT})} \quad . \end{aligned} \tag{3.48}$$

Here, σ_W is used as the parameter which indicates the emission homogeneity of the emitter. The smaller the σ_W value the more homogeneous is the emission. If $\sigma_W \rightarrow 0$, the space charge limited region and the temperature limited region of the CVC curve will be connected at a single point, as shown in the blue line of Figure 3.1.

Equations 3.48 and 3.28 can be used as the fitting function for the experimental current voltage data [ATS05]. During the fitting process, the work function distribution σ_W is an unknown parameter. The other parameters (e.g., T, W_0, $\beta_{\text{C-L}}$ andβ_E) can be set to known values or unknown fitting variables. More the known parameters are given in the fitting process, the more accurate σ_W will be.

3.2.3 New definition of emission inhomogeneity

Both the Khodnevich and the Tonnerre model have their own definition of the emission uni- formity, but none of them gives the current density inhomogeneity directly. The Khodnevich model neglects the Schottky effect and gives the effective emission inhomogeneity. The Tonnerre model gives the emission inhomogeneity in the form of the work function distribution σ_W, but for this model, σ_W is very sensitive to the given value of temperature T and average work function W_0. Correct values of temperature T and W_0 have to be known in order to get the correct σ_W value. However, W_0 is a function of T [Cro81]. Due to the complex and compact geometry of a MIG, it is difficult to know the correct emitter temperature. In this work, the most intuitive definition of emission inhomogeneity is used, which is given by

$$\delta_J = \frac{\sigma_J}{\overline{J}} \tag{3.49}$$

where σ_J is the standard deviation of the current density and $\overline{J}$ is the average value of the current density. To calculate δ_J, the equation 3.48 is used. Since both the Khodnevich and the Tonnerre model need a second derivative of the CVC curve, they need a very high accuracy of the experimental data. A fitting method will be more robust. The test process for the calculation of δ_J is shown in Figure 3.6.

After the measurement of the CVC curve, a fitting is made using equation 3.48. During the fitting process, the average work function is a fixed value while T, $\beta_{\text{C-L}}$, β_E and σ_W are free parameters. Then, a list of work functions that have a Gaussian distribution $N(W_0, \sigma_W)$ are generated using a random number generator. The parts of the emitter whose work function W_i is smaller than W_t works under the space charge limited region; the left parts of the emitter work under the temperature limited region. $J_{\text{C-L}}$ and J_{S} are given by equations 3.1 and 3.3, respectively. After the calculation of J_i for each part of the emitter, δ_J can be derived.

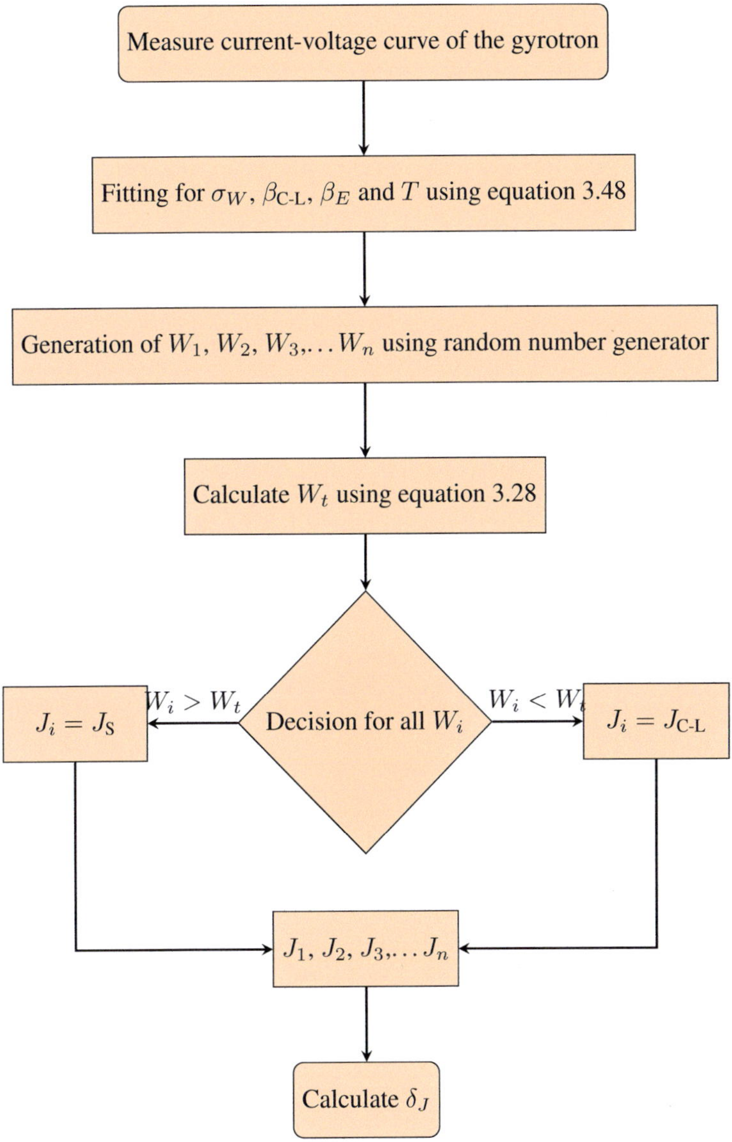

Figure 3.6: Calculation process of normalized current density distribution.

As discussed above, the uncertainty of work function W and temperature T will influence the accuracy of σ_W. To show the advantage of the new definition of emission inhomogeneity, a comparison on the influence of work function is made between different definitions. Table 3.1 gives the comparison of the emission inhomogeneity with different definition of one typical CVC curve. The average work function W_0 varies from 1.7 eV to 2.0 eV, as shown in the first column of Table 3.1. The influence of the uncertainty of W_0 on the work function distribution σ_W and emitter temperature are shown in the second and third column.

Table 3.1: Comparison of the emission inhomogeneity with different definition.

W_0 [eV]	σ_W [eV]	T [K]	δ_J
1.7	0.0236	1087.35	0.2558
1.8	0.0248	1144.75	0.2558
1.9	0.0261	1201.95	0.2558
2.0	0.0273	1258.85	0.2559

In Table 3.1, σ_W changes with the given work function W_0, while δ_J has a constant value. Therefore, the new definition of the emission inhomogeneity δ_J is more stable. In this work, the intuitive and stable parameter δ_J with the calculation method described in Figure 3.6 is used to calculate the emission inhomogeneity.

3.3 Investigation of experimental results

The KIT 15 kW, 28 GHz gyrotron for materials processing [MIP+13], the EU 1 MW, 170 GHz short-pulse prototype gyrotron for ITER [JAA+13, PAA+15] and the experimental first industrial prototype of EU 2 MW,

170 GHz coaxial cavity gyrotron [PDD+04, RPK+10] have been measured and analyzed in this work. The 15 kW, 28 GHz gyrotron has been operated for several months. The 1 MW, 140 GHz gyrotron for W7-X is a new gyrotron tube which has just gone under testing. The EU 2 MW, 170 GHz coaxial cavity gyrotron has already been in operation since several years and the emitter has been exposed to air several times. The experimental data for the 28 GHz gyrotron are measured according to an automatic current-voltage measurement system. The experimental data for the W7-X and coaxial gyrotron are measured point by point due to the limitation of the high voltage power supply.

3.3.1 Experimental results for the KIT 15 kW, 28 GHz gyrotron

Firstly, the experimental results of the KIT 15 kW, 28 GHz gyrotron are analyzed using the three methods. The measured CVCs are shown in Figure 3.7.

The different colours of lines and points in Figure 3.7 represent different emitter filament current (emitter temperature). The dots represent experimental data while the lines are the calculated results from using the CVC fitting equation. The calculated results for the KIT 15 kW, 28 GHz gyrotron are shown in Table 3.2. During the calculation $W_0 = 1.85\,\mathrm{eV}$ is provided by the manufacturer of the emitter.

In Table 3.2, T and σ_W are the results obtained using the fitting method. δ_J is the calculated current emission inhomogeneity using the new definition, which is described in equation 3.49. A more detailed information of emission inhomogeneity for this gyrotron can be derived using the Tonnerre model. The calculated work function distribution $D(W)$ using equations 3.39 and 3.44 of curve "T1" for the KIT 28 GHz gyrotron is shown in Figure 3.8.

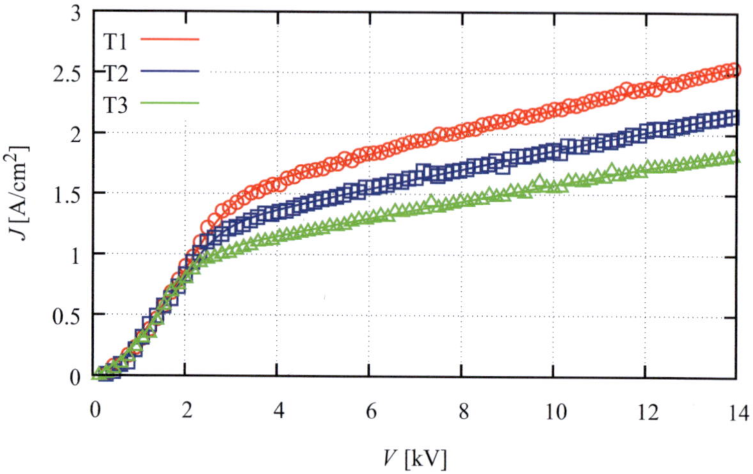

Figure 3.7: Current voltage characteristics of the KIT 15 kW, 28 GHz gyrotron for three different emitter temperatures. T1=1172.95 K, T2=1162.85 K, T3=1152.05 K.

Table 3.2: Calculation results of 28 GHz gyrotron for different emitter temperature.

Curve	T [K]	σ_W [eV]	δ_J
T1	1172.95	0.0248	0.249
T2	1162.85	0.0255	0.259
T3	1152.05	0.0269	0.276

One can see from Figure 3.8 that the results of these two methods are quite similar to each other. There is a shift of the average work function from the given value of about 0.01 eV. It is assumed that this shift results from the different operating temperatures compared to the testing temperature of the manufacturer. The Tonnerre model provides more detailed information about the work function distribution. For example, in the red line of

Figure 3.8, there are two small peaks before the main peak, which indicate that there are two high current density areas on the emitter surface.

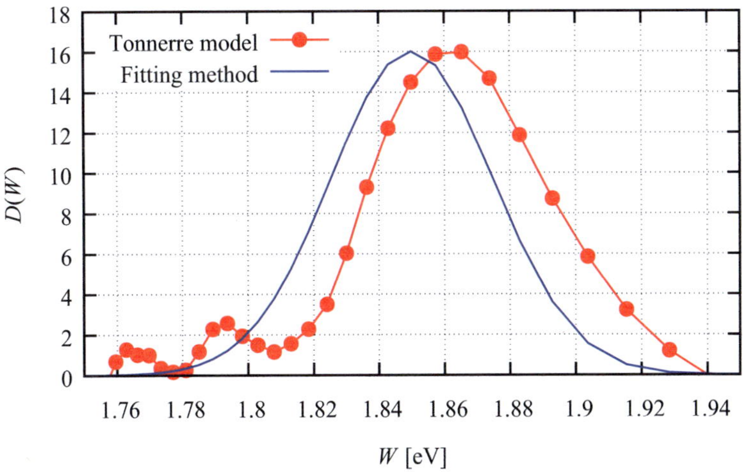

Figure 3.8: Work function distribution of the KIT 28 GHz gyrotron.

3.3.2 Experimental results for the EU 1 MW, 170 GHz ITER gyrotron

Figure 3.9 and Table 3.3 present the experimental results for the EU 1 MW, 170 GHz ITER gyrotron with different emitter temperature. During the calculation, $W_0 = 2.1\,\text{eV}$ is provided by the manufacturer of the emitter. The curve "T1" in Figure 3.9 is the CVC of the gyrotron under the nominal cathode working temperature. As shown in Table 3.3, this emitter has excellent emission homogeneity under the nominal working temperature compared to the KIT 15 kW, 28 GHz gyrotron. The corresponding emission inhomogeneity δ_J is only 0.048, which is less than one-quarter of the KIT 15 kW, 28 GHz gyrotron.

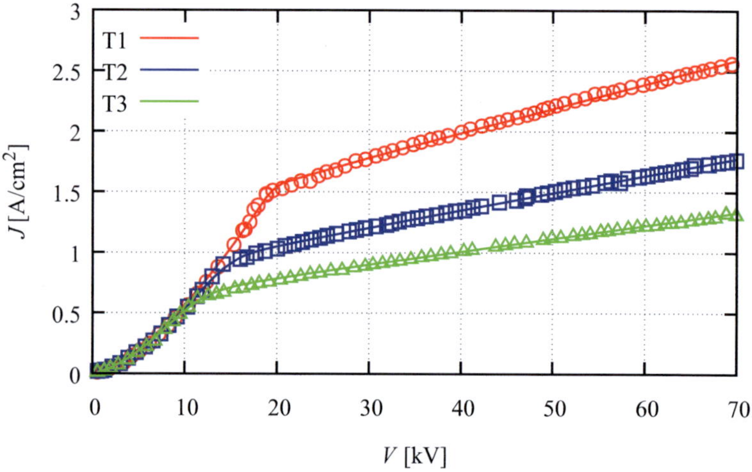

Figure 3.9: Current voltage characteristics of the EU 1 MW, 170 GHz ITER gyrotron for three different emitter temperatures. T1=1310.65 K, T2=1284.15 K, T3=1264.85 K.

Table 3.3: Calculated results for the EU 1 MW, 170 GHz ITER gyrotron for three different emitter temperatures.

Curve	T [K]	σ_W [eV]	δ_J
T1	1310.65	0.0054	0.048
T2	1284.15	0.0175	0.159
T3	1264.85	0.0274	0.256

3.3.3 Experimental results for the first short-pulse prototype of the EU 2 MW, 170 GHz coaxial cavity gyrotron

Figure 3.10 and Table 3.4 are the experimental and fitted results for the EU 2 MW, 170 GHz coaxial cavity short-pulse prototype gyrotron. One can see from Figure 3.10 that the transition from the space charge limited region to the temperature limited region is much wider than for the other two gyrotrons. This indicates that the cathode of the coaxial gyrotron has a higher emission inhomogeneity. The calculated results in Table 3.4 confirm that the inhomogeneity of the coaxial-cavity gyrotron is much higher than that of the other two gyrotrons, which indicates that the emission inhomogeneity is increasing with the increase of operating time and number of tube openings.

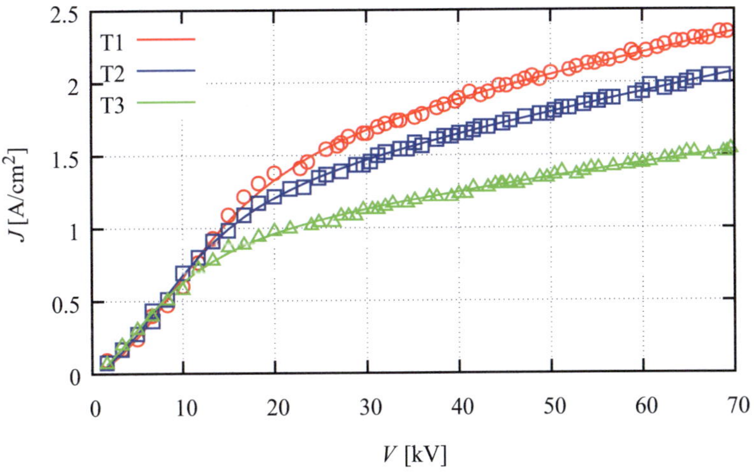

Figure 3.10: Current voltage characteristics of the first short-pulse prototype of the EU 2 MW, 170 GHz coaxial cavity gyrotron for three different emitter temperatures. T1=1308.65 K, T2=1294.85 K, T3=1273.25 K.

Table 3.4: Calculation results of the first short-pulse prototype of the EU 2 MW, 170 GHz coaxial cavity gyrotron for three different emitter temperatures.

Curve	T [K]	σ_W [eV]	δ_J
T1	1308.65	0.0680	0.661
T2	1294.85	0.0752	0.760
T3	1273.25	0.0943	1.043

From the experimental results of these three gyrotrons, one can see that δ_J has the minimum value at the working temperature, and will increase with the decrease in temperature. Two factors can cause this phenomenon. Firstly, during the design of the emitter surface, the uniformity of the electric field is higher at the working temperature, and will decrease with the change in temperature because of the deformation of the cathode surface. Increased electric field inhomogeneity at a lower temperature can increase the emission inhomogeneity. Secondly, the thermionic emitters used in this work are dispenser cathodes, which are made of porous tungsten impregnated with barium oxide. The ratio and distribution of barium oxide on the cathode surface can be different at different temperatures. This difference could also result in the increase of emission inhomogeneity.

3.4 Influences on electron beam quality and gyrotron efficiency

To study the influence of emission inhomogeneity the 3D trajectory code ARIADNE [PV04] was used. A Gaussian distribution for the work function on the emitter surface was assumed. Figure 3.11 is the current density on the emitter surface. The current density is distributed symmetrically on the emitter surface.

In reality, the current density distribution around the emitter surface can be arbitrary. In order to simplify the calculation, the distribution of current density shown in Figure 3.11 is used. Influence of emission inhomogeneity on the KIT 15 kW, 28 GHz gyrotron and the EU 1 MW, 170 GHz ITER gyrotron were calculated and compared in this work.

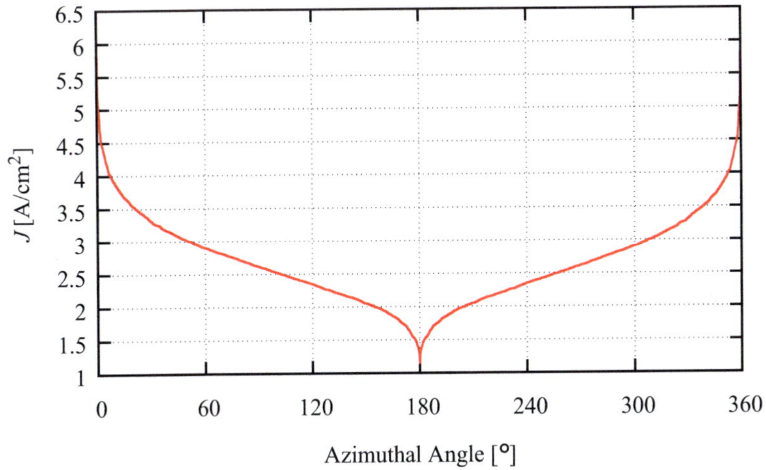

Figure 3.11: 2D plot of the current density along the emitter azimuthal angle.

3.4.1 Influence of emission inhomogeneity on the beam quality of the KIT 15 kW, 28 GHz gyrotron

The combined influence of emission inhomogeneity and surface roughness on the beam quality at the centre of the cavity is shown in Table 3.5.

In Table 3.5, $\bar{\alpha}$ and α_{max} are the average and maximum value of the pitch factor α, respectively. One can see from Table 3.5 that both emission inhomogeneity and emitter surface roughness will increase α_{max} and the velocity spread (δ_α, $\delta_{\beta_\perp}$). $\bar{\alpha}$ is 1.634 without the influence of emission inhomogeneity and surface roughness. In this case, trapped electrons can be generated when the emission inhomogeneity δ_J is 0.238 and surface roughness size r_0 is larger than $5\,\mu\mathrm{m}$. Trapped electrons can be generated immediately after the emission inhomogeneity δ_J increases to 0.5. The influence of trapped electrons will be discussed in Chapters 5 and 6.

Table 3.5: Combined influence of emission inhomogeneity and surface roughness on beam quality of the KIT 15 kW, 28 GHz gyrotron.

σ_W [eV]	δ_J	r_0 [μm]	$\bar{\alpha}$	α_{max}	$\delta_{\beta_\perp}$ [%]	$\delta_{E_{\mathrm{kin}}}$ [%]	$\delta_{r_{\mathrm{gc}}}$ [%]
0.0	0.0	0	1.634	2.00	2.53	0.409	4.26
		3	1.665	2.80	4.95	0.458	4.40
		5	1.673	3.00	5.28	0.462	4.43
0.025	0.238	0	1.662	2.70	3.18	0.616	6.44
		3	1.670	3.65	5.20	0.634	6.57
		5	Trapped electrons!				
0.05	0.5	0					

A detailed information about the influence of emission inhomogeneity on the electron beam guiding centre and energy distribution at the centre of the cavity are shown in Figure 3.12.The red points are the calculated results for a homogeneous emitter while the blue points are the calculated results for an inhomogeneous emitter ($\delta_J = 0.238$). The x-axis is the azimuthal angle of the cylindrical electron beam. Figure 3.12(a) is the guiding centre distribution of the electrons along the azimuthal angle. For the homogeneous electron beam (red points) the electrons are distributed uniformly along the azimuthal angle. For the inhomogeneous electron beam (blue points), there is a very strong jitter of the guiding centre along the azimuthal angle. The maximum guiding centre difference of the inhomogeneous electron beam is 1.8 times larger than for the homogeneous electron beam. The guiding centre spread $\delta_{r_{\mathrm{gc}}}$ is increased by 2.2 % (from

4.26 % to 6.44 %), because of the emission inhomogeneity. As shown in Figure 3.12(b), the same behaviour happens on the energy spread. On the other hand, the energy spread δ_E is only increased by 0.2 % (from 0.409 % to 0.616 %) due to the emission inhomogeneity.

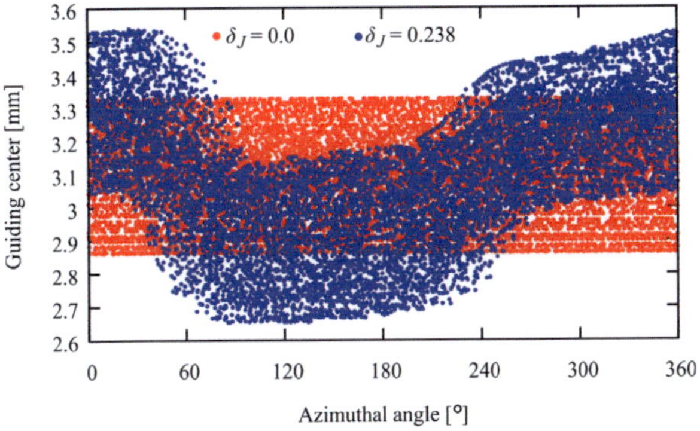

(a) Guiding centre distribution.

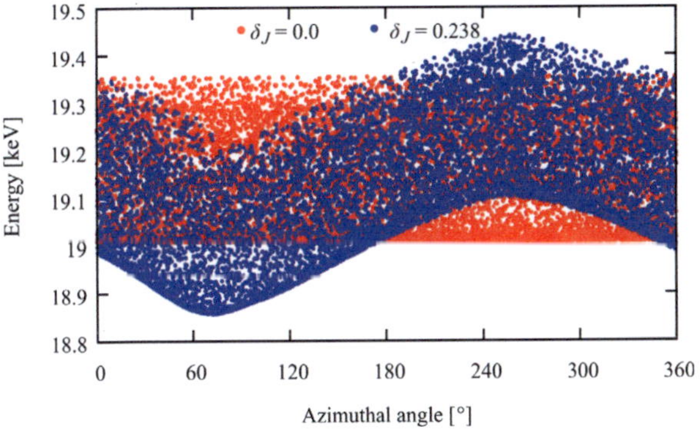

(b) Energy distribution.

Figure 3.12: Influence of emission inhomogeneity on the guiding centre (a) and energy distribution (b) at the centre of the cavity for the KIT 15 kW, 28 GHz gyrotron.

3.4.2 Influence of emission inhomogeneity on the beam quality of the EU 1 MW, 170 GHz ITER gyrotron

The combined influence of emission inhomogeneity and surface roughness on the beam quality of the 170 GHz 1 MW MIG at the centre of the cavity is shown in Table 3.6.

Table 3.6: Combined influence of emission inhomogeneity and surface roughness on beam quality of the EU 1 MW, 170 GHz ITER gyrotron.

σ_W [eV]	δ_J	r_0 [µm]	$\bar{\alpha}$	α_{max}	$\delta_{\beta_\perp}$ [%]	$\delta_{E_{\text{kin}}}$ [%]	$\delta_{r_{gc}}$ [%]
0.0	0.0	0	1.50	1.74	2.61	1.019	1.226
		3	1.52	2.84	4.20	1.023	1.227
		5	1.54	3.39	4.83	1.023	1.227
0.025	0.223	0	1.50	1.78	2.85	1.805	1.260
		3	1.52	3.26	4.41	1.822	1.259
		5	1.53	3.45	4.96	1.826	1.260
0.05	0.464	0	1.49	1.84	3.57	3.064	1.360
		3	1.52	3.58	5.00	3.143	1.365
		5	1.53	3.91	5.52	3.200	1.371

Similar to the KIT 15 kW, 28 GHz gyrotron, Table 3.6 shows that both the emission inhomogeneity and the emitter surface roughness will increase α_{max} and the velocity spread (δ_α, $\delta_{\beta_\perp}$) for the EU 1 MW, 170 GHz ITER gyrotron. However, for the average pitch factor $\bar{\alpha} = 1.50$ used in the simulation, no trapped electrons are generated up to $\delta_J = 0.464$ and $r_0 = 5\,\mu\text{m}$. These last two columns show that surface roughness has a very limited influence on the kinetic energy and guiding centre spread compared to the influence of emission inhomogeneity.

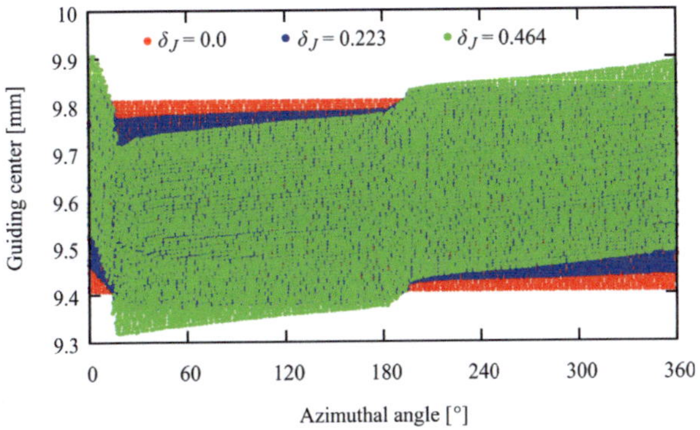

(a) Guiding centre distribution.

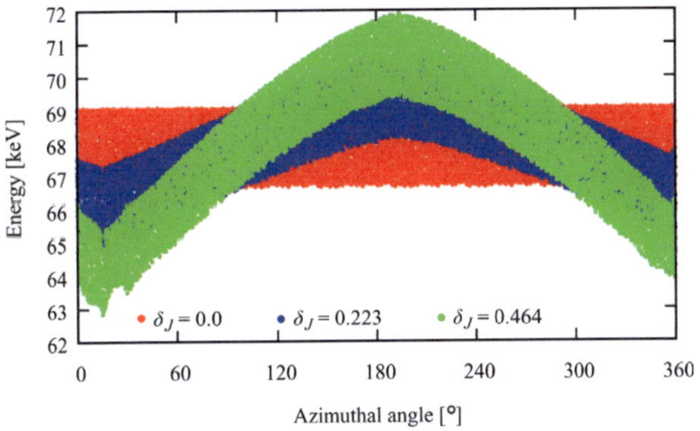

(b) Energy distribution.

Figure 3.13: Influence of emission inhomogeneity on the (a) guiding centre and (b) energy distribution at the centre of the cavity for the EU 1 MW, 170 GHz ITER gyrotron.

Figure 3.13 shows the influence of emission inhomogeneity on guiding centre and energy distribution at the centre of the cavity. One can see from Figures 3.13 and 3.12 that the influence of the emission inhomogeneity on guiding centre distribution for EU 1 MW, 170 GHz ITER gyrotron is not as strong as it is for the KIT 15 kW, 28 GHz gyrotron. For the 170 GHz gyrotron $\delta_{r_{\mathrm{gc}}}$ is only increased by 0.034 % for the emission inhomogeneity $\delta_J = 0.223$, while for the 28 GHz gyrotron $\delta_{r_{\mathrm{gc}}}$ is increased by 2.2 % for the emission inhomogeneity $\delta_J = 0.238$. This is because the magnetic field in the 170 GHz gyrotron (6.7 T) is much stronger than the magnetic field in the 28 GHz gyrotron (0.5 T). However, the influence on energy spread $\delta_{E_{\mathrm{kin}}}$ of the electrons for the 170 GHz gyrotron is much stronger than for the 28 GHz gyrotron. For the same emission inhomogeneity, $\delta_{E_{\mathrm{kin}}}$ is increased by 0.8 % and 0.2 % for the 170 GHz and 28 GHz gyrotron, respectively. This is because the current is much higher in the 170 GHz MIG (40.0 A) than in the 28 GHz MIG (2.0 A).

3.4.3 Influence of emission inhomogeneity on the gyrotron efficiency

The influence of emission inhomogeneity on the efficiency of the KIT 15 kW, 28 GHz gyrotron and the EU 1 MW, 170 GHz ITER gyrotron are calculated using the code EURIDICE. Electrons used in the calculation of EURIDICE are imported from the 3D trajectory code ARIADNE. The calculation results of the influence of emission inhomogeneity on gyrotron efficiency for these two gyrotrons are shown in Figure 3.14.

To avoid trapped electrons which can not be handled by the trajectory code the average pitch factor α for the KIT 15 kW, 28 GHz gyrotron is decreased to 1.48 in the calculation. The red points are calculation result for the KIT 15 kW, 28 GHz gyrotron and the blue points are for the EU 1 MW, 170 GHz ITER gyrotron. The influence of surface roughness is not considered in this calculation.

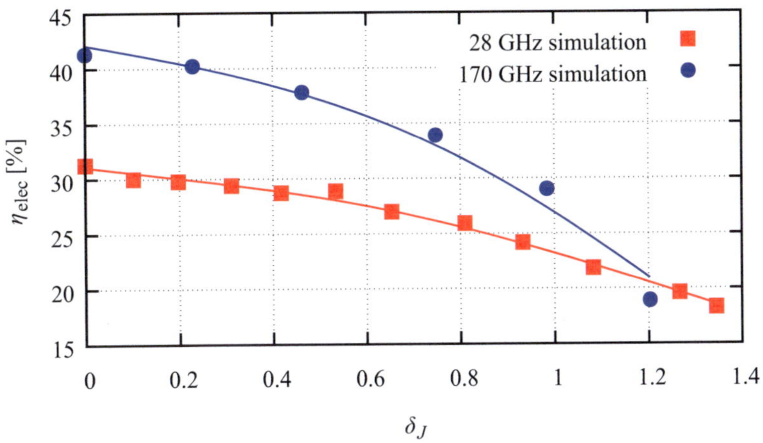

Figure 3.14: Influence of emission inhomogeneity on gyrotron efficiency.

As shown in Figure 3.14, emission inhomogeneity decreases the efficiency for both gyrotrons. When the emission inhomogeneity δ_J increases from 0.4 to 1.1, the efficiency of the 170 GHz gyrotron decreases by approximately 15 %, which is comparable to the 14 % decrease of the experimental result from [GGK+99]. For both the 170 GHz and 28 GHz gyrotron, the emission inhomogeneity δ_J =0.3 results in 2 % decrease of the efficiency, so that δ_J =0.3 is defined as the upper limit for the emission inhomogeneity of the gyrotron emitter.

3.4.4 Upper limits for the work function, temperature and electric field distribution

According to equation 1.3, the emission inhomogeneity is influenced by the distribution of work function, electric field, and temperature. Taking the temperature distribution as an example, its influence on the emission inhomogeneity can be calculated using the method shown in the flowchart plotted in Figure 3.15.

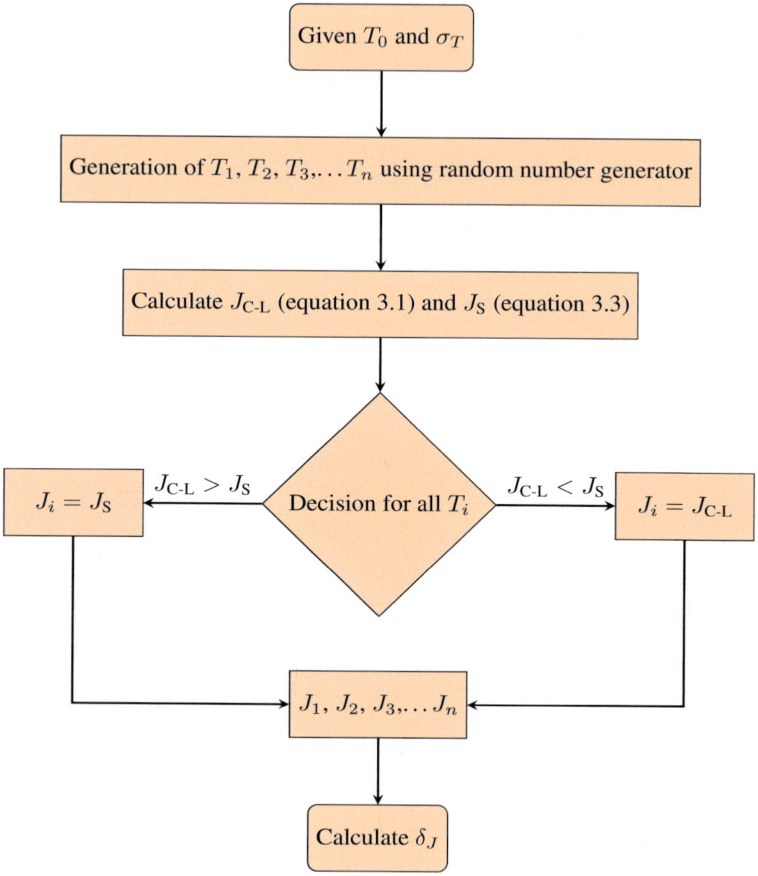

Figure 3.15: Calculation process for the influence of temperature distribution σ_T on the emission inhomogeneity δ_J.

Firstly, the average temperature T_0 and standard deviation σ_T are given. Then, a list of temperature values which have a Gaussian distribution

$N(T_0, \sigma_T)$ are generated using a random number generator. Each of the temperature value T_i is corresponding to a small fraction of the emitter surface with the same area. Subsequently, the space charge limited current density $J_{\text{C-L}}$ and the temperature limited current density J_{S} for each fraction of the emitter are calculated using equation 3.1 and 3.3, respectively. For each fraction of the emitter, if the space charge limited current density $J_{\text{C-L}}$ is smaller than the temperature limited current density J_{S}, it works under the space charge limited region. Its corresponding current density $J_i = J_{\text{C-L}}$. Otherwise, it works under temperature limited region and $J_i = J_{\text{S}}$. After J_i for each fraction of the emitter is calculated, the emission inhomogeneity δ_J can be derived. The influence of the work function and the electric field distribution on the emission inhomogeneity can be calculated similar to the above method. Influence of the temperature, work function and electric field distribution on the emission inhomogeneity of four different gyrotrons are shown in Figure 3.16.

For the influence of electric field distribution on the emission inhomogeneity δ_E is used as the evaluation factor for the distribution. The definition of δ_E is given by

$$\delta_E = \frac{\sigma_E}{E_0} \quad . \tag{3.50}$$

Here, σ_E and E_0 are the standard deviation and average of the electric field, respectively. These four gyrotrons used in the calculation of Figure 3.16 are: the KIT 15 kW, 28 GHz gyrotron, the SN-5i 1 MW, 140 GHz gyrotron for the experimental stellarator Wendelstein 7-X [GEI+10], the EU 1 MW, 170 GHz ITER gyrotron, and the first short-pulse prototype of the EU 2 MW, 170 GHz coaxial cavity gyrotron. The distribution of temperature, work function, and electric field have a similar influence on the emission inhomogeneity of these four gyrotrons.

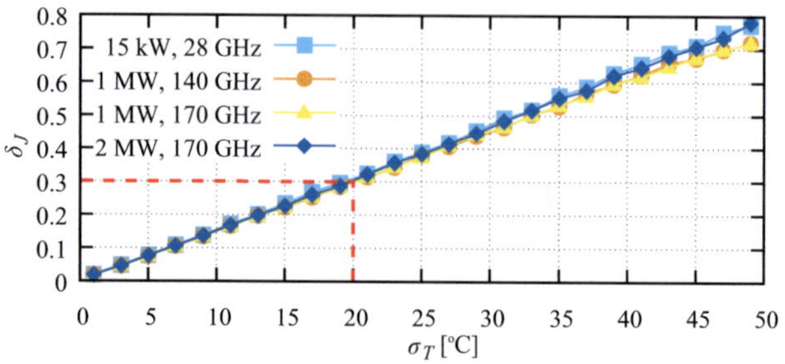

(a) Influence of temperature distribution on emission inhomogeneity.

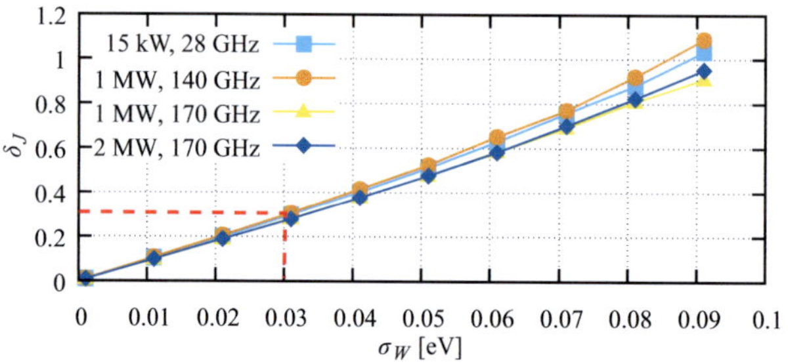

(b) Influence of work function distribution on emission inhomogeneity.

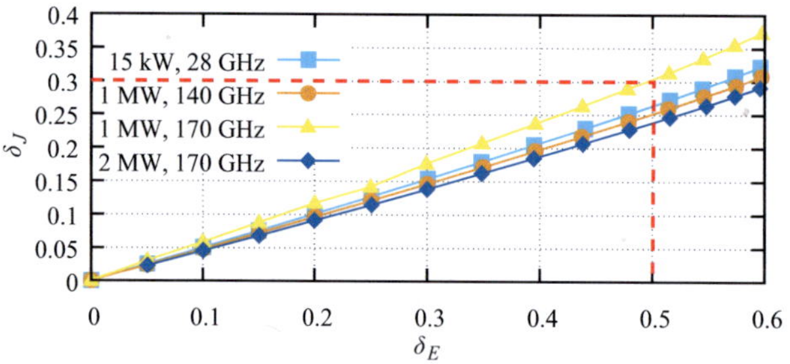

(c) Influence of electric field distribution on emission inhomogeneity.

Figure 3.16: Influence of the temperature, work function and electric field distribution on the emission inhomogeneity of four different gyrotrons.

According to the requirement in Section 3.4.3, to keep δ_J smaller than 0.3 the corresponding separate upper limits for the distribution of work function, temperature and electric field are

$$\begin{aligned} \sigma_T &< 20\,^{\circ}\mathrm{C} \\ \sigma_W &< 0.03\,\mathrm{eV} \\ \delta_E &< 0.5 \end{aligned} \tag{3.51}$$

It shall be pointed out that the emission inhomogeneity is a compound result of the temperature, work function, and electric field distribution. Therefore, the real separate upper limits for each parameter will be lower. For new emitters, the emission inhomogeneity are affected more by temperature variations than work function variations [IBC+08]. During the cathode aging process, the emission inhomogeneity will be increased and it is affected more by work function variations instead of temperature variations [Lon03]. For modern new emitters the requirement for $\delta_J < 0.3$ is always fulfilled (e.g., emitters for the KIT 15 kW, 28 GHz gyrotron and the EU 1 MW, 170 GHz ITER gyrotron). The emitter ageing process during the gyrotron operation can be monitored by the method derived in this chapter.

3.5 Summary

Two methods for the determination of the emission inhomogeneity of gyrotron MIG emitters based on the current voltage characteristics have been presented and modified. A new definition of emission inhomogeneity and its evaluation method is derived based on the two presented methods. The new definition is intuitive and delivers more stable results. With the help of the new evaluation method, the experiment and analysis results of three different gyrotron MIGs are presented. The comparison of those three gyrotron MIGs indicates that the emission inhomogeneity probability will increase with the operation time and number of tube openings. The influence of emission inhomogeneity on electron beam quality and gyrotron efficiency is calculated using the 3D code ARIADNE and EURIDICE. The numerical investigations show that $\delta_J = 0.3$ will decrease the gyrotron efficiency by approximately $2\,\%$. To have sufficient emission homogeneity the distribution of temperature σ_T, work function σ_W and electric field δ_E should be smaller than $20\,^{\circ}$C, 0.03 eV and 0.5, respectively. Numerical simulation using ARIADNE also shows that emission inhomogeneity can result in trapped electrons in the gyrotron. Investigations on the influence of trapped electrons in the gyrotron will be presented in Chapter 5 and Chapter 6.

4 Secondary electron emission model for magnetron injection gun

The investigation on surface roughness and emission inhomogeneity described in the previous Chapters 2 and 3 shows that these two adverse factors can result in trapped electrons in the gyrotron. Secondary electrons can be generated by the bombardment of the cathode surface by back reflected trapped electrons. To study the influence of trapped electrons and generated secondary electrons (see Chapters 5 and 6), a modified secondary electron emission model for gyrotron MIGs is developed in this chapter.

After an introduction to the Furman secondary electron emission model [FP02], the angular distributions of elastically reflected and rediffused secondary electrons are calculated using the Monte Carlo code CASINO [DPDC$^+$11]. Then, the calculation algorithm of the modified secondary electron emission model is described. At the end of this chapter, the secondary emission yields of tungsten and molybdenum, which are used in the cathode, are calculated. This secondary emission model can also be used for calculations on the collector where copper is used.

4.1 Secondary electron emission model

When an electron is trapped by the magnetic mirror (as introduced in Section 1.4.1 and in more detail in Section 5.1), it can be back reflected along the magnetic field line and then bombards the cathode surface. The bombardment generates secondary electrons, which have high initial kinetic energy (from several eV to several keV). Secondary electrons with high initial kinetic energy have a higher probability of being trapped by the magnetic mirror again. The accumulation of trapped and secondary electrons in the gyrotron MIG can cause Low Frequency Oscillations (LFOs, Chapter 5) and an Electron Beam Halos (EBHs, Chapter 6) in the gyrotron.

According to the former research on surface roughness (Chapter 2), the electron beam quality is quite sensitive to the initial velocity of emitted electrons, therefore, the Furman secondary electron emission model [FP02], which has detailed description on secondary electron distributions is chosen as the basic model for this work.

4.1.1 Furman secondary electron emission model

According to [FP02], if an electron beam bombards a material surface, three types of secondary electrons will be emitted from the solid surface: elastic electrons, rediffused electrons, and true secondary electrons. Figure 4.1 shows the generation process of these three types of secondary electrons.

I_0 is the current of the initially injected electron beam. The electrons which are reflected back by the outmost layer of atoms are called elastic electrons and I_e is the current of these elastically backscattered electrons. The electrons which enter the material and are scattered back by the inner atoms are called rediffused electrons; the resulting current is I_r. The true secondary electrons are emitted from the atoms during the collision of the incident electrons with the atoms and are finally scattered out from the material. The current of the true secondary electrons is I_{ts}. Finally, the Secondary Emission Yield (SEY) is defined as

$$\begin{aligned}\delta &= \frac{I_e + I_r + I_{ts}}{I_0} \\ &= \delta_e + \delta_r + \delta_{ts} \quad . \end{aligned} \tag{4.1}$$

Here, δ_e, δ_r and δ_{ts} are the secondary electron emission yield factors of elastic, rediffused, and true secondary electrons, respectively.

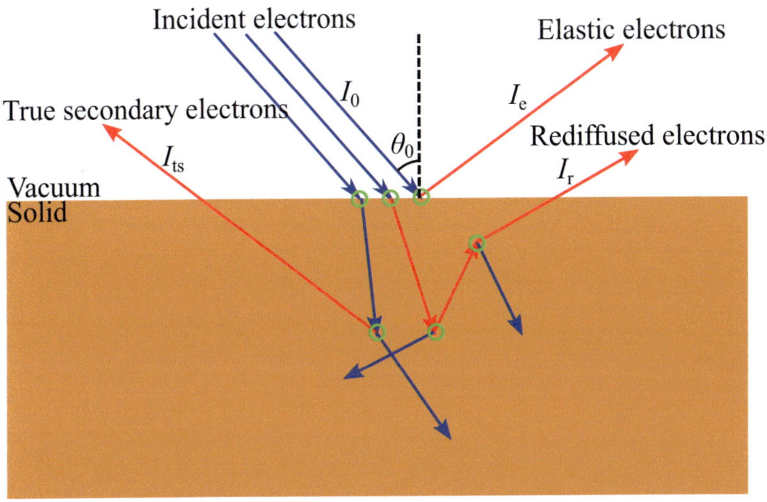

Figure 4.1: Sketch of secondary electron emission.

The SEY, energy, and angular distribution are the three most important parameters for the secondary electron emission model. Figure 4.2 shows the energy spectrum for secondary electrons emitted from a copper surface when the electron beam is injected with 100 eV initial energy. The direction of the electron beam is perpendicular to the sample surface. For this case, $\delta_e = 0.112$, $\delta_r = 0.179$, $\delta_{ts} = 1.401$. 82.8 % of the secondary electrons are true secondary electrons, these electrons usually have several eV initial energy (not higher than 50 eV). 10.5 % of the secondary electrons are rediffused secondary electrons, these electrons have a very large energy region from several eV to the initial injection energy. 6.7 % of the secondary electrons are elastic secondary electrons, these electrons usually have a very high initial energy (nearly the same as the initial injection energy). According to the research on surface roughness which uses the initial velocity model, very small initial energy of the electrons emitted from the emitter can result in a significant increase of the velocity spread of the electron beam in the MIG. Thus, it is necessary to consider all of these three types of secondary electrons.

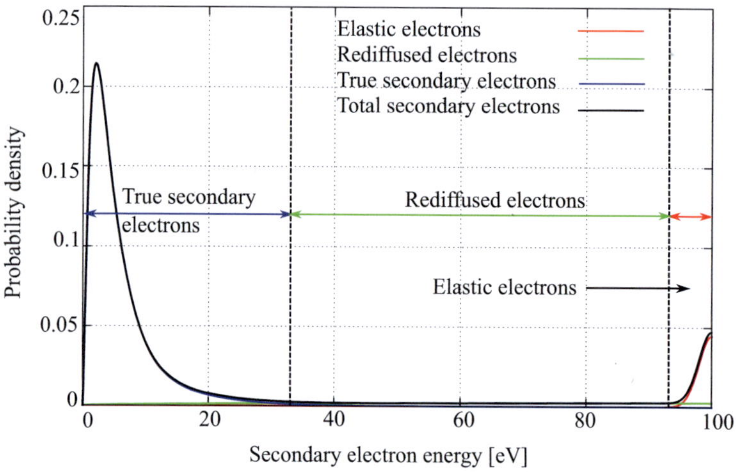

Figure 4.2: Secondary electron energy spectrum of copper ($E_{\mathrm{kin},0} = 100\,\mathrm{eV}$, $\theta_0 = 0\,°$).

In the Furman secondary electron emission model, the SEY and emitted-energy spectrum are provided with a broad phenomenological fit to experimental data using different equations. The following three subsections describe the electron yield and energy spectrum for the three types of secondary electrons, respectively.

Model for elastic electrons

According to [FP02], the elastic electron yield is given by

$$\delta_e(E_{\text{kin},0}, 0) = P_e(\infty) + (\hat{P}_e - P_e(\infty))e^{-(|E_{\text{kin},0} - \hat{E}_e|/W)^p/p} \quad (4.2)$$

$$\delta_e(E_{\text{kin},0}, \theta_0) = \delta_e(E_{\text{kin},0}, 0)[1 + e_1(1 - \cos^{e_2}\theta_0)] \quad . \quad (4.3)$$

Here, $\delta_e(E_{\text{kin},0}, 0)$ is the elastic electron yield for the injection angle $\theta_0 = 0\,°$ (as is shown in Figure 4.1) and $E_{\text{kin},0}$ is the initial injection energy, $\hat{P}_e$ is the peak value of $\delta_e(E_{\text{kin},0}, 0)$ when $E_{\text{kin},0} = \hat{E}_e$ and $P_e(\infty)$ is the value of $\delta_e(E_{\text{kin},0}, 0)$ when $E_{\text{kin},0} = \infty$. p, e_1 and e_2 are variables in the equation which will be defined according to the fit of equations 4.2 and 4.2 to the experimental data.

In Furman secondary emission model, the energy spectrum of the elastic electrons is given by

$$f_e = \chi(E_{\text{kin}})\chi(E_{\text{kin},0} - E_{\text{kin}})\delta_e(E_{\text{kin},0}, \theta_0)\frac{2e^{-(E_{\text{kin}} - E_{\text{kin},0})^2/2\sigma_e^2}}{\sqrt{2\pi}\sigma_e \text{erf}(E_{\text{kin},0}/\sqrt{2}\sigma_e)} \quad . \quad (4.4)$$

Here, $\chi(x)$ is the indicator function which is equal to 1 when $x \geq 0$, otherwise the value is 0. With $\chi(x)$ it is guaranteed that the kinetic energy of elastic electrons follows: $0 < E_{\text{kin}} < E_{\text{kin},0}$. erf(x) is the error function, σ_e is a variable which can be derived by fitting to experimental data.

Model for rediffused electrons

The yield and energy spectrum of rediffused electrons in the Furman secondary emission model are given by

$$\delta_r(E_{\text{kin},0},0) = P_r(\infty)[1-e^{-(E_{\text{kin},0}/E_{\text{kin},r})^r}] \quad (4.5)$$

$$\delta_r(E_{\text{kin},0},\theta_0) = \delta_r(E_{\text{kin},0},0)[1+r_1(1-\cos^{r_2}\theta_0)] \quad . \quad (4.6)$$

Here, $\delta_r(E_{\text{kin},0},0)$ is the elastic electron yield for the injection angle $\theta_0 = 0\,^\circ$ (as is shown in Figure 4.1). $P_r(\infty)$ is the value of $\delta_r(E_{\text{kin},0},0)$ when $E_{\text{kin},0} = \infty$. $E_{\text{kin},r}$, r, r_1, and r_2 are variables in the equation which will be defined according to the fit of equation 4.5, equation 4.6, and equation 4.7 to the experimental data.

The energy spectrum of the rediffused electrons is given by

$$f_r = \chi(E)\chi(E_{\text{kin},0}-E)\delta_r(E_{\text{kin},0},\theta_0)\frac{(q+1)E^q}{E_{\text{kin},0}^{q+1}} \quad . \quad (4.7)$$

q is also an variable similar to $E_{\text{kin},r}$, r, r_1 and r_2 which will be defined according to the fit of equation 4.5, equation 4.6 and equation 4.7 to the experimental data.

Model for true secondary electrons

In the Furman secondary emission model, the relation between the secondary emission yield of the true secondary electrons δ_{ts}, the incident electron energy and the angle of incidence is

$$\delta_{ts}(E_{\text{kin},0},\theta_0) = \hat{\delta}(\theta_0)D(E_{\text{kin},0}/E_{\text{kin}}(\theta_0)) \quad , \quad (4.8)$$

where

$$\begin{aligned} D(x) &= \frac{sx}{s-1+x^s} \\ \hat{\delta}(\theta_0) &= \hat{\delta}_{ts}[1+t_1(1-\cos^{t_2}\theta_0)] \\ E_{\text{kin}}(\theta_0) &= E_{ts}[1+t_3(1-\cos^{t_4}\theta_0)] \quad . \end{aligned}$$

In equation 4.8, δ_{ts}, E_{ts}, s, t_1, t_2, t_3, and t_4 are variables in the equation, which will be defined according to the fit of equation 4.8 to the experimental data.

The energy spectrum of true secondary electrons follows as [PEP+02]

$$f_{ts} = \frac{\frac{E}{E_{pk}} e^{\frac{-E}{E_{pk}}}}{E_{pk} - (E_{\text{kin},0} + E_{pk}) e^{-\frac{E_{\text{kin},0}}{E_{pk}}}} \quad . \tag{4.9}$$

E_{pk} is the empirical energy where the energy spectrum function of the true secondary electrons has its maximum.

Probability model of the number of secondary electrons

The probability of the number of secondary electrons generated per electron in the Furman model is given by

$$\begin{aligned} P_0 &= (1 - \delta_e - \delta_r) P_{0,ts} \\ P_1 &= (1 - \delta_e - \delta_r) P_{1,ts} + \delta_e + \delta_r \\ P_n &= (1 - \delta_e - \delta_r) P_{n,ts} \end{aligned} \quad . \tag{4.10}$$

P_n is the probability to generate n secondary electrons and $P_{n,ts}$ is the probability to generate n true secondary electrons. When $n = 0$, no secondary electron or true secondary electron will be generated. $P_{n,ts}$ is a binomial distribution in the Furman model, which is given by

$$P_{n,ts} = \begin{pmatrix} M \\ n \end{pmatrix} p^n (1 - p)^{M-n} \quad . \tag{4.11}$$

Here, $p = \delta'_{ts}/M$, $\delta'_{ts} = \delta_{ts}/(1 - \delta_e - \delta_r)$ and M is the maximum number of secondary electrons that can be generated by one incident electron. Therefore, the probabilities satisfy

$$\sum_{n=0}^{M} P_n = 1 \tag{4.12}$$

$$\sum_{n=0}^{M} P_{n,ts} = 1 \quad . \tag{4.13}$$

According to the above definition of the probability to generate n secondary electrons, the average number of secondary electrons generated per bombardment, or the SEY satisfies

$$\delta = \sum_{n=0}^{M} nP_n = \delta_{ts} + \delta_e + \delta_r \quad . \tag{4.14}$$

4.1.2 Angular distribution of secondary electrons

The angular distribution of secondary electrons is another important parameter for the secondary electron emission model. The definition of angles is shown in Figure 4.3. The electron beam is incident in the yz-plane; the xy-plane defines the interface between vacuum and solid.

In the Furman model, it is assumed that all the emission angles of the secondary electrons have the same angular distribution which is given by [SM71, SIM72]

$$f_{ts}(\theta) = \sin(2\theta) \tag{4.15}$$

where θ represents the emission angles of the secondary electron under consideration.

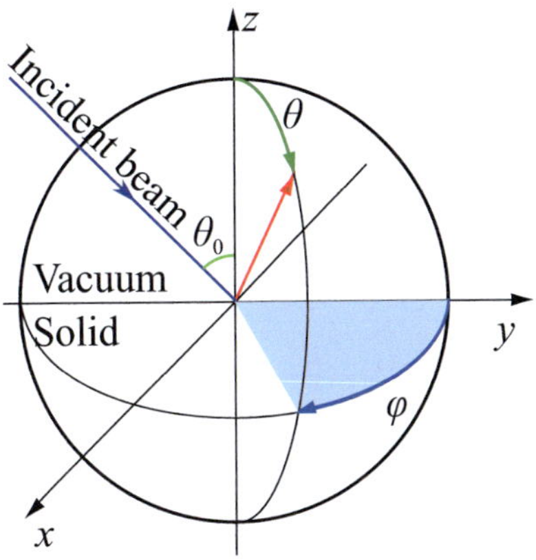

Figure 4.3: Spherical coordinates of Monte Carlo calculation result.

The angular distribution of the secondary electrons shown in equation 4.15 is independent of the primary incidence angle θ_0 and incident energy E_0. This is not quite true for the elastically reflected and rediffused electrons, which are scattered out by the atoms. The elastically reflected and rediffused electrons also obey the above $\sin(2\theta)$ distribution [Kan57, SM71, MMS71, SIM72] for normal incident and small inclined incidence, as is shown in Figure 4.4. The red circles in Figure 4.4 are the angular probability that the elastically reflected and re-diffused electrons are scattered into the solid angle. The distribution (red circles) will change with the incident electron angle and energy, as shown in Figure 4.4 for a highly inclined incidence.

To calculate the angular distribution for the elastically reflected and rediffused electrons the Monte Carlo code CASINO [DPDC+11] is used. CASINO includes an accurate model for electron microscopy applications. The original intent of this software is to assist scanning electron microscope users in interpretation of imaging and microanalysis [DCJ+07]. It can calculate the relation between the primary incidence angle θ_0, incident energy E_0, and the angular distribution of the elastically reflected and rediffused electrons.

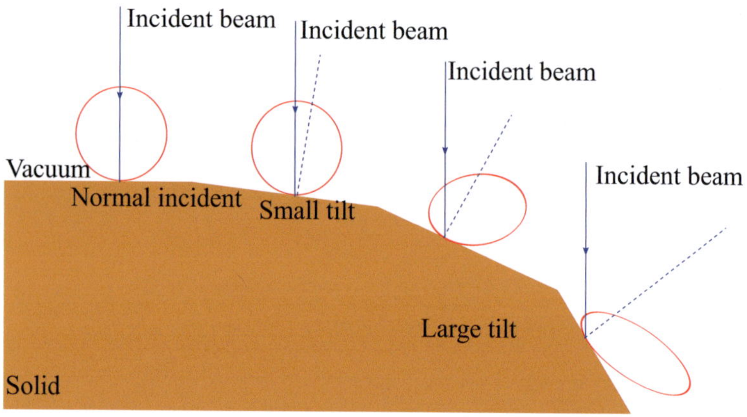

Figure 4.4: Angular distribution of elastic and rediffused electrons.

At first, the situation is assumed when copper is bombarded by an 1 keV electron beam. According to the calculation result obtained from CASINO, the relation between the incidence angle θ_0 and the outward angle θ and φ for the elastically reflected and rediffused electrons is shown in Figure 4.5.

The different line points stand for different incidence angle θ_0. Figure 4.5 shows that when the incidence angle is less than 60°, the θ distribution of the elastic and rediffused electrons obey the cosine law and the φ distribution is nearly uniform. When the incidence angle increases to a value larger than 60°, both the θ and φ distributions change significantly.

Besides the incidence angle, the angular distribution of the elastically reflected and rediffused electrons depends also on the initial energy of the incident electrons and the bombarded type of material. To see the influence of the energy of the incident electron beam, an electron beam with an 10 keV initial energy is calculated and the comparison of the result with that for a 1 keV incident electron beam is shown in Figure 4.6.

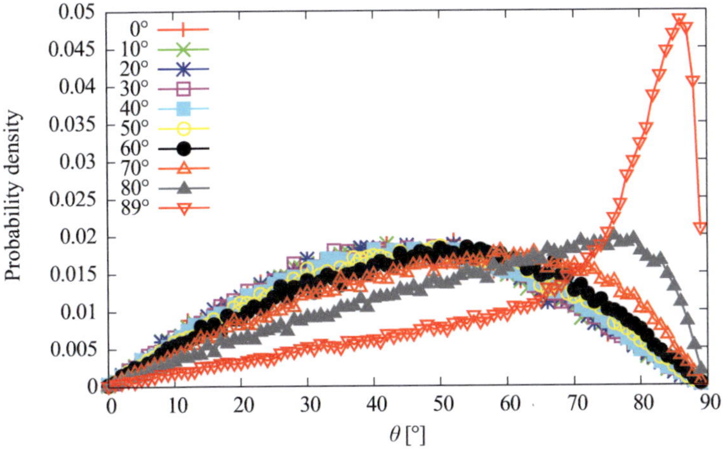

(a) θ distribution of the elastically reflected and rediffused electrons

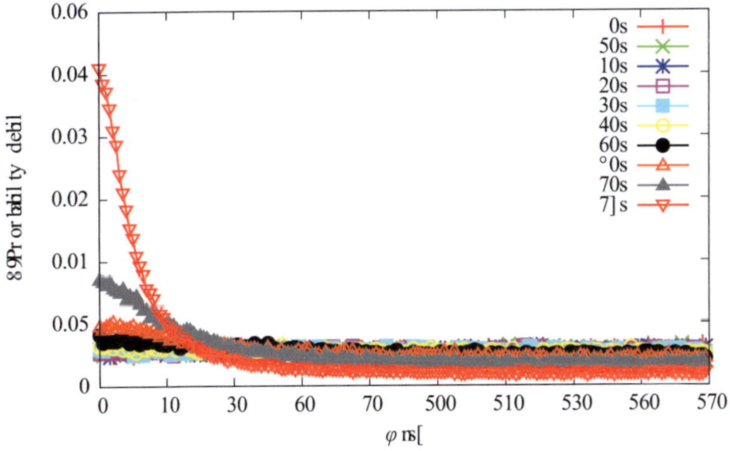

(b) φ distribution of the elastically reflected and rediffused electrons

Figure 4.5: Relation between the incidence angle θ_0 and the outward angle θ and φ for the elastically reflected and rediffused electrons of copper.

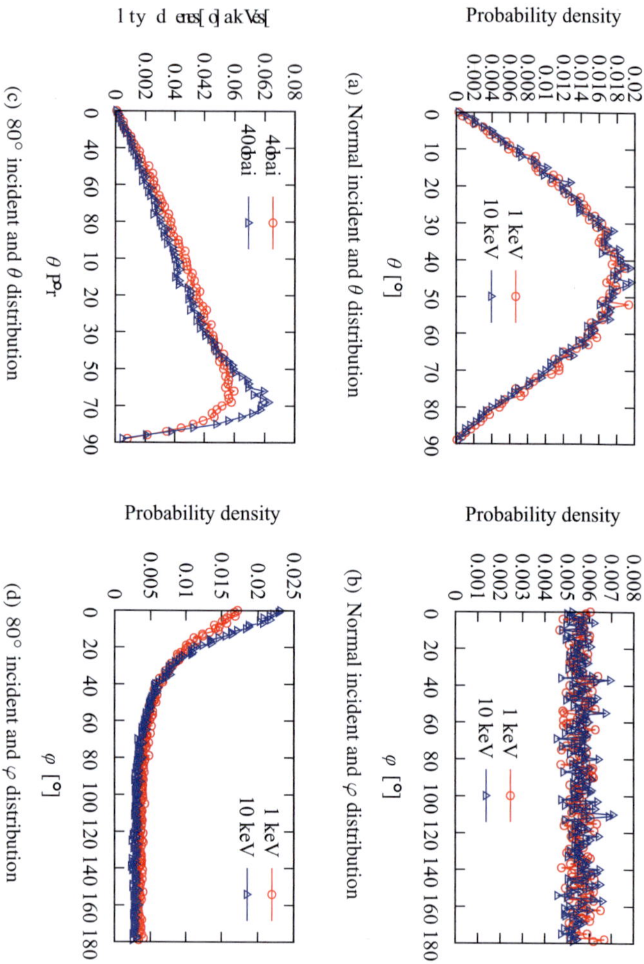

Figure 4.6: Influence of the incident energy on the angular distribution of elastically reflected and rediffused electrons. (a) and (b): for incidence angle $\theta_0 = 0°$, (c) and (d): for incidence angle $\theta_0 = 80°$.

Figures 4.6(a) and 4.6(b) show the angular distribution of the elastic and rediffused secondary electrons when the beam is incident normally to the solid surface. Figures 4.6(c) and 4.6(d) are the angular distribution of the elastic and rediffused secondary electrons when the incident beam angle θ_0 is 80°. When the incidence angle is normal to the solid surface the angular distributions of the elastic and rediffused electrons is the same for the 1 keV and 10 keV incident beam. When the incidence angle θ_0 increases to 80°, the influence of the energy is still not significant. Therefore, in the model, the influence of initial incidence energy on the angular distribution of the elastic and rediffused electrons is neglected.

Besides the initial incidence energy, the influence of different types of material is also compared. There are three kinds of materials that may be bombarded by the electron beam in a gyrotron: copper, Stainless Steel (SS), and molybdenum. To see the influence of the different materials on the angular distribution, the bombarding of the incident electron beam with different initial incident energies and angles on these three materials are calculated separately. Figure 4.7 shows the results when the initial beam is bombarding the solid surface with different energy and incidence angles.

Figures 4.7(a) and 4.7(b) show the angular distributions of the elastic and rediffused secondary electrons when the incident angle of the electron beam is normal to the surface of materials with 1 keV initial energy. Figures 4.7(c) and 4.7(d) show the angular distribution of the elastic and rediffused secondary electrons when the incident beam angle is 80° and the initial beam energy is 1 keV. Figure 4.7(e) and Figure 4.7(f) are the angular distribution of the elastic and rediffused secondary electrons when the incident beam angle is 80° and the initial beam energy is 10 keV. From Figure 4.7, one can see that the influence of the material is quite small; therefore, the influence of the material is neglected in the model.

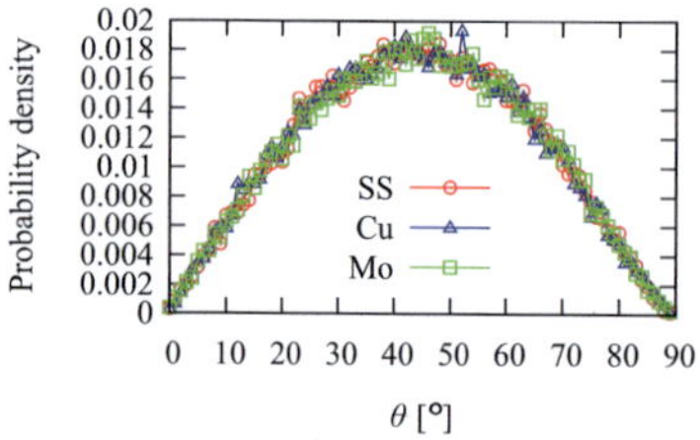

(a) Normal incident of 1 keV electron beam

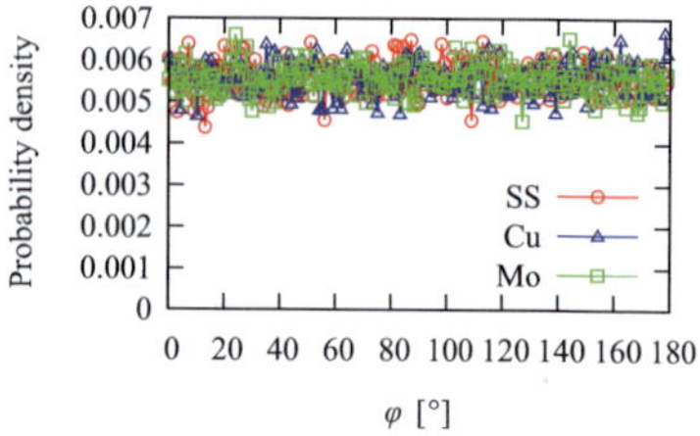

(b) Normal incident of 1 keV electron beam

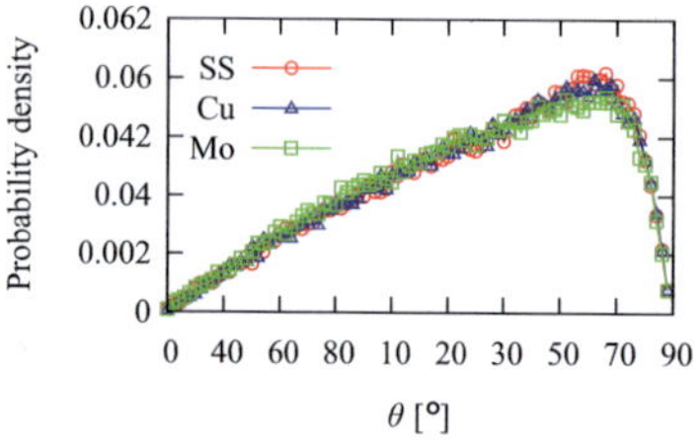

(c) 80° incident of 1 keV electron beam

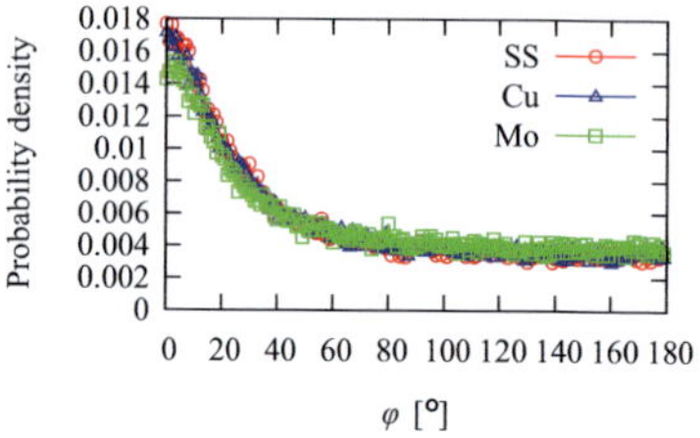

(d) 80° incident of 1 keV electron beam

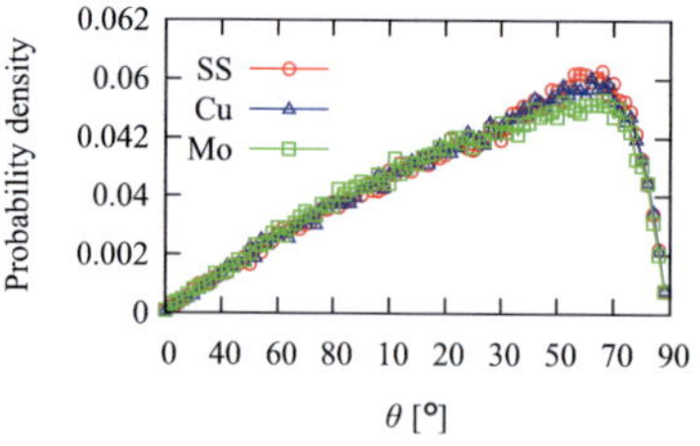

(e) 80° incident of 10 keV electron beam

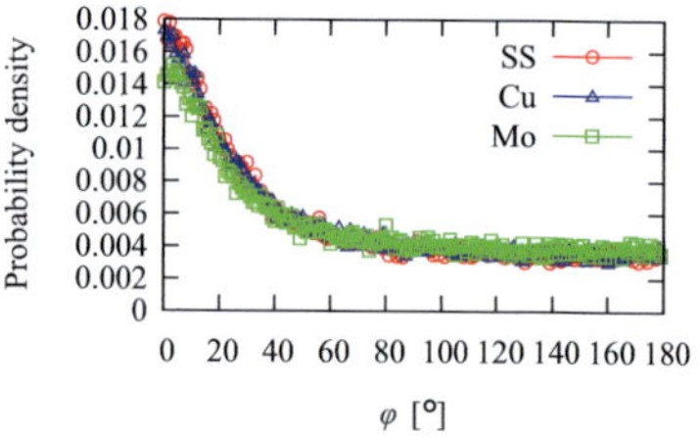

(f) 80° incident of 10 keV electron beam

Figure 4.7: Influence of the material on the angular distribution of the elastically reflected and rediffused electrons.

According to the calculation and analysis described above, the angular distribution of the elastic and rediffused electrons is only related to the incidence angle θ_0 of the initial electron beam. Therefore, the cases of incident electron beam with different θ_0 are calculated and analyzed based on this assumption. The calculation results are shown in Figure 4.8.

It is shown that as the incidence angle increases, an increasing number of elastic and rediffused electrons have a high θ and low φ angle. This behavior is not included in the Furman model. To include this behaviour, a polynomial surface fit of up to 9×9 degrees was made using MATLAB. The fit function can be described as given in equation 4.16.

$$\begin{aligned} f_{e,r}(\theta, \theta_0) &= \sum_{i=0}^{i=9} \sum_{j=0}^{j=9-i} p_{ij} \theta^i \theta_0^j \\ &= \boldsymbol{\theta} \boldsymbol{P}_{coef} \boldsymbol{\theta_0}^{\mathrm{T}} \end{aligned} \tag{4.16}$$

with:

$$\boldsymbol{\theta} = \begin{bmatrix} 1 & \theta & \theta^2 & \theta^3 & \theta^4 & \theta^5 & \theta^6 & \theta^7 & \theta^8 & \theta^9 \end{bmatrix},$$

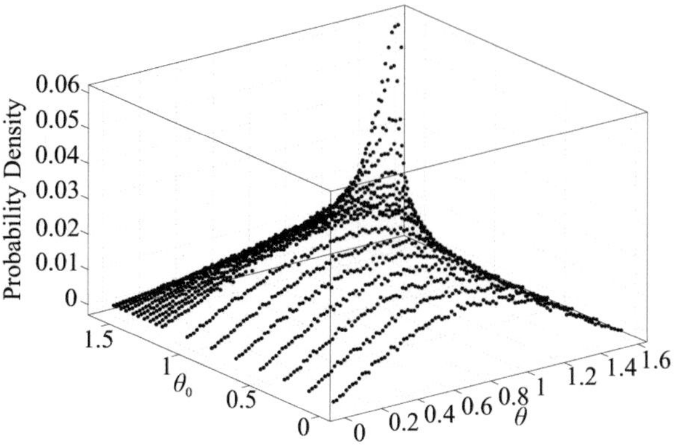

(a) θ distribution

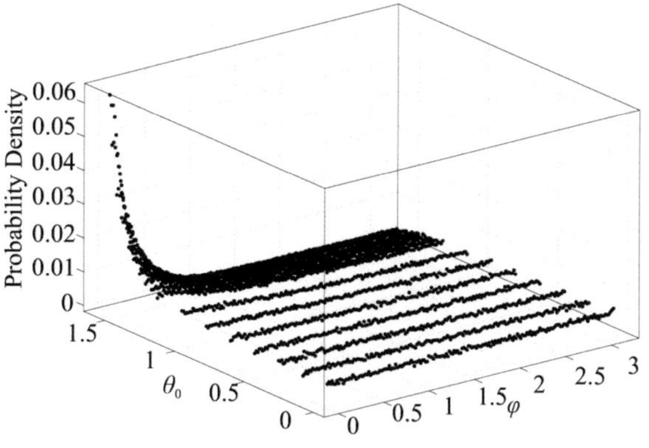

(b) φ distribution

Figure 4.8: Angular distribution of elastic and rediffused electrons with different θ_0 obtained from CASINO.

$$
\boldsymbol{P}_{coef} = \begin{bmatrix}
p_{00} & p_{01} & p_{02} & p_{03} & p_{04} & p_{05} & p_{06} & p_{07} & p_{08} & p_{09} \\
p_{10} & p_{11} & p_{12} & p_{13} & p_{14} & p_{15} & p_{16} & p_{17} & p_{18} & 0 \\
p_{20} & p_{21} & p_{22} & p_{23} & p_{24} & p_{25} & p_{26} & p_{27} & 0 & 0 \\
p_{30} & p_{31} & p_{32} & p_{33} & p_{34} & p_{35} & p_{36} & 0 & 0 & 0 \\
p_{40} & p_{41} & p_{42} & p_{43} & p_{44} & p_{45} & 0 & 0 & 0 & 0 \\
p_{50} & p_{51} & p_{52} & p_{53} & p_{54} & 0 & 0 & 0 & 0 & 0 \\
p_{60} & p_{61} & p_{62} & p_{63} & 0 & 0 & 0 & 0 & 0 & 0 \\
p_{70} & p_{71} & p_{72} & 0 & 0 & 0 & 0 & 0 & 0 & 0 \\
p_{80} & p_{81} & 0 & 0 & 0 & 0 & 0 & 0 & 0 & 0 \\
p_{90} & 0 & 0 & 0 & 0 & 0 & 0 & 0 & 0 & 0
\end{bmatrix},
$$

$$
\boldsymbol{\theta_0} = \begin{bmatrix} 1 & \theta_0 & \theta_0^2 & \theta_0^3 & \theta_0^4 & \theta_0^5 & \theta_0^6 & \theta_0^7 & \theta_0^8 & \theta_0^9 \end{bmatrix}.
$$

Here, the middle matrix is defined to be $\boldsymbol{P}_{coef}$, which means it is the coefficient matrix between $\boldsymbol{\theta}$ and $\boldsymbol{\theta_0}$. $f_{e,r}(\theta, \theta_0)$ gives the θ distribution of the elastic and rediffused electrons under any incidence angle θ_0. The same fitting is made for the φ distribution which gives $g_{e,r}(\theta, \theta_0)$. $f_{e,r}(\theta, \theta_0)$ and $g_{e,r}(\theta, \theta_0)$ are included in the codes ESRAY and ESPIC. Therefore, the code can calculate secondary electrons with higher accuracy of angular distribution compared to the Furman model.

4.2 Calculation algorithm and results

As discussed in Section 4.1, the SEY, the energy, and angular distributions are the three most important parameters for the description of secondary electrons. In the code ESRAY and ESPIC, when a trapped electron is back reflected and bombards the cathode surface, the SEY and energy of the secondary electrons are calculated using the Furman model which is introduced in Section 4.1.1. The angular distribution of the secondary electrons is calculated using $f_{e,r}(\theta, \theta_0)$ and $g_{e,r}(\theta, \theta_0)$, which are derived in Section 4.1.2. In the following subsection, the calculation algorithm for the generation of secondary electrons in the code ESRAY and ESPIC is described.

4.2.1 Calculation algorithm

The flowchart of the secondary electron emission model for the codes ESRAY and ESPIC is shown in Figure 4.9.

Step 1. During gyrotron MIG calculations, parts of electrons are reflected by the magnetic mirror. When the electron bombards the cathode surface, the incidence energy $E_{\text{kin},0}$ and the incidence angle θ_0 of the electron are calculated.

Step 2. The SEY, $\delta_e(E_{\text{kin},0}, \theta_0)$, $\delta_r(E_{\text{kin},0}, \theta_0)$ and $\delta_{ts}(E_{\text{kin},0}, \theta_0)$ according to equations 4.3, 4.6 and 4.8 are defined.

Step 3. The number of secondary electrons n generated by the incident electron is calculated according to equations 4.10 and 4.11.

Step 4. If $n = 0$, no secondary electron is generated and the MIG calculation continues.

Step 5. If $n = 1$, calculate the type of the secondary electron according to $\delta_e(E_{\text{kin},0}, \theta_0)$, $\delta_r(E_{\text{kin},0}, \theta_0)$, and $\delta_{ts}(E_{\text{kin},0}, \theta_0)$ using a uniform random number generator.

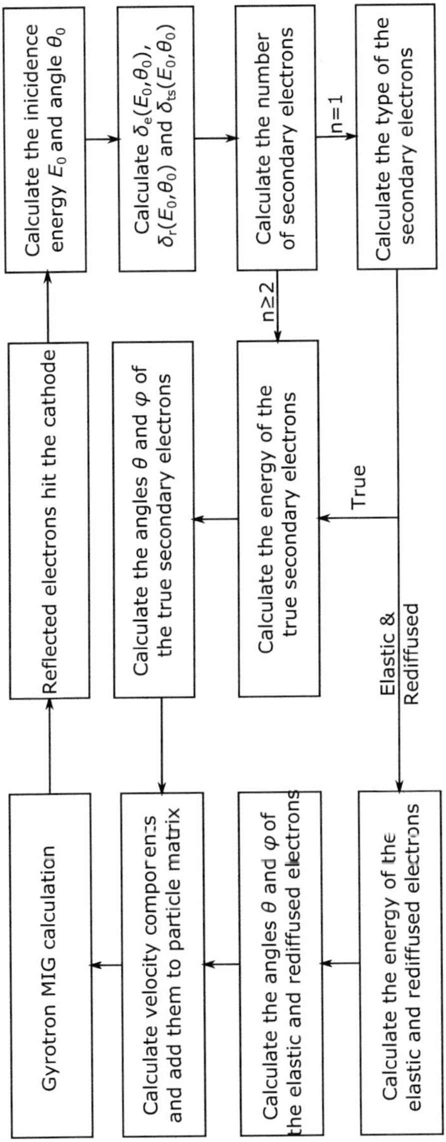

Figure 4.9: Calculation algorithm of secondary electron.

(a) If $0 \leq$ random(1) $\leq \delta_e(E_{\text{kin},0}, \theta_0)$, the secondary electron is an elastic electron.

(b) If $\delta_e(E_{\text{kin},0}, \theta_0) \leq$ random(1) $\leq \delta_e(E_{\text{kin},0}, \theta_0)+\delta_r(E_{\text{kin},0}, \theta_0)$, the secondary electron is a rediffused electron.

(c) Else, it is a true secondary electron.

Step 6. If $n \geq 2$, all of them are true secondary electrons.

Step 7. The energy of the secondary electron is calculated according to the type of the electron using equation 4.4, equation 4.7 and equation 4.9.

Step 8. For elastic and rediffused electrons, the angle θ and φ are calculated according to the polynomial fit (equation 4.16).

Step 9. For true secondary electrons, the angle is calculated according to equation 4.15. Then, a random φ value between $[-\pi, \pi]$ which shall have a uniform distribution, is given to the secondary electron.

Step 10. The velocity components for each secondary electron are calculated according to its energy E, angle θ and φ. Add all information to the particle matrix.

Step 11. The gun calculation continues.

4.2.2 Calculation results

The SEY, the energy spectrum and the energy spectrum of secondary electrons are calculated according to the described calculation algorithm.

Secondary electron emission yield

Figure 4.10 shows the SEY of Cu when $0\,\text{eV} \leq E_{\text{kin},0} \leq 2000\,\text{eV}$ and θ_0 is $0°$; the secondary emission parameters are given by [FP02]. It is shown that the SEY increases with the incident energy at the beginning and then decreases after the SEY reaches the maximum. The initial increase of the SEY is due to the fact that with the increase in incident energy more and more true secondary electrons are generated in the material. Therefore, the number of secondary electrons that can get out of the material will increase. At the same time, electrons will go deeper into the material which will make it harder for these true secondary electrons to get out. After a certain point, even more true secondary electrons are generated in the solid material but the number of true secondary electrons that can get out of the material will decrease.

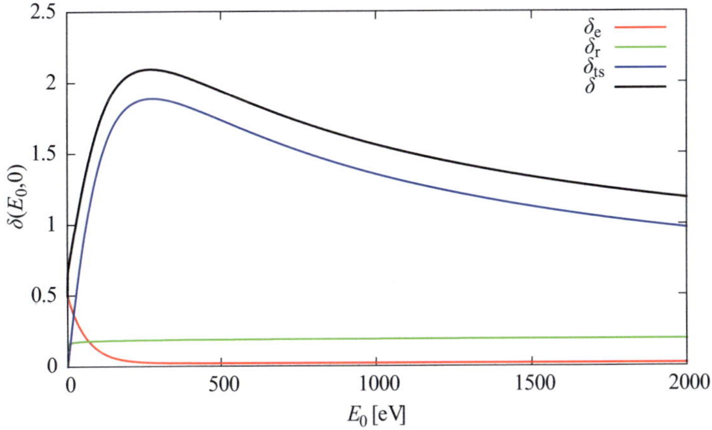

Figure 4.10: Calculated SEY of copper ($0\,\text{eV} \leq E_{\text{kin},0} \leq 2000\,\text{eV}$, $\theta_0 = 0°$).

Figure 4.11 shows the calculation result on the relation between SEY and the incidence angle θ_0 when the incident energy $E_{\text{kin},0}$ is $300\,\text{eV}$. $E_{\text{kin},0} =$

300 eV is chosen for this calculation, because both copper and stainless steel have a very high SEY value at this point and the influence of θ_0 on SEY will be more significant.

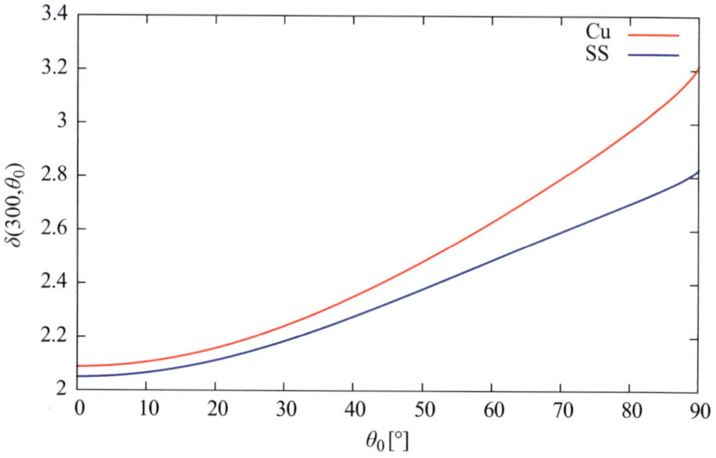

Figure 4.11: Relation between the incidence angle θ_0 and SEY of copper and stainless steel ($E_{\text{kin},0} = 300$ eV).

It is shown that with the increase of incidence angle θ_0, there is a constant increase of the SEY. This is because the increase of the incidence angle results in a decrease in the distance of the secondary electron to the material surface. Therefore, the generated secondary electrons in the material can more easily get out of the solid.

Energy spectrum

Figure 4.12 shows the secondary electron energy spectrum when a 300 eV electron beam bombards copper and stainless steel surfaces. It is shown that copper creates a larger number of low energy secondary electrons (0-20 eV) and a smaller number of high energy secondary electrons (approximately

300 eV). This result shows that different materials have different secondary electron energy spectrums. According to the results for surface roughness, electrons with high initial energy will increase the velocity spread and have a higher probability of being trapped in the MIG. These trapped electrons will increase the number of secondary electrons in turn. Thus, a material with lower SEY and lower probability of elastic and rediffused electron should be chosen as the cathode material.

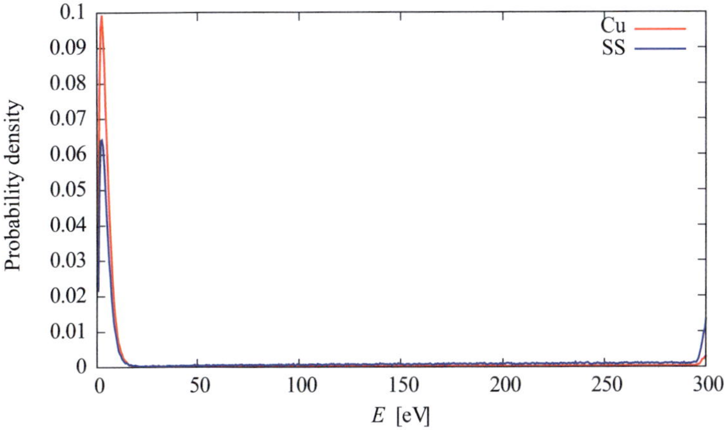

Figure 4.12: Secondary electron energy spectrum of copper and stainless steel.

Angular distribution

Figure 4.13 shows the angular distribution of secondary electrons when the initial electron beam energy $E_{\text{kin},0}$ is 300 eV and the incidence angle is 80°. The red line is the angular distribution of all the secondary electrons and the blue line is the angular distribution of the elastic and rediffused electrons. It can be seen that more elastic and rediffused electrons have high θ and low φ value like in Figure 4.8.

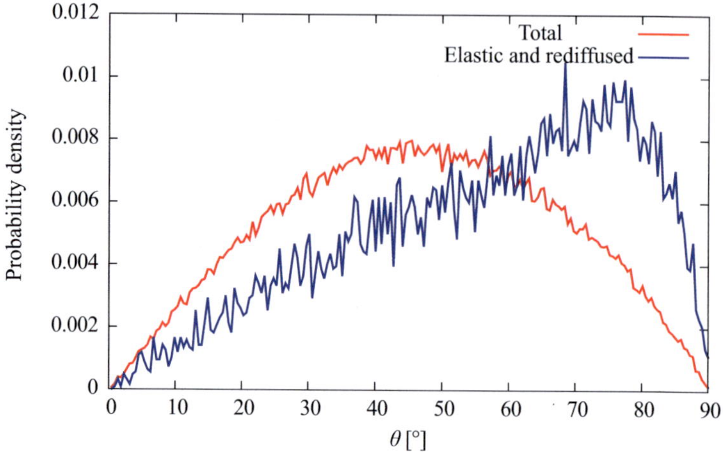

(a) θ distribution of secondary electrons

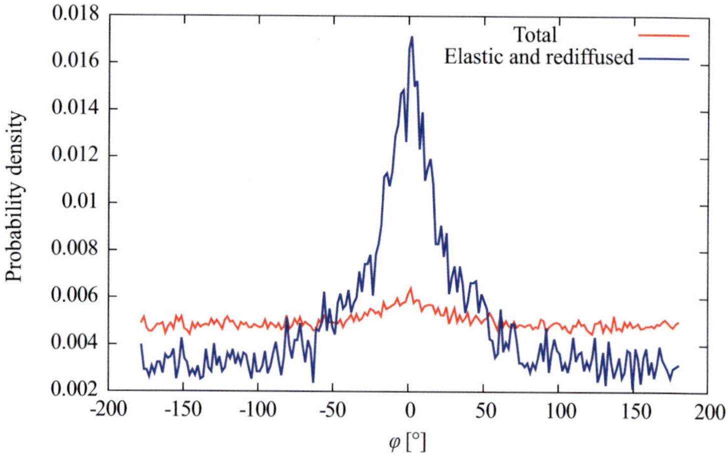

(b) φ distribution of secondary electrons

Figure 4.13: Calculated angular distribution of secondary electrons. $E_{\mathrm{kin},0} = 300\,\mathrm{eV}$, $\theta_0 = 80^\circ$.

4.3 Secondary electron emission properties of W and Mo

Copper is used in the collector of the gyrotron while tungsten and molybdenum are two widely used materials for the cathode. Tungsten is the main material used for the emitter ring; molybdenum is usually used for the cathode subcomponents on both sides of the emitter ring. In the Furman secondary emission model, the secondary electron emission parameters are only given for copper and stainless steel. In this work, the secondary electron emission parameters for tungsten and molybdenum are deduced by fitting the existing experimental data [Joy95] using equation 4.2, equation 4.5 and equation 4.8. The secondary emission yields of tungsten and molybdenum are shown in Figure 4.14.

The gray points are secondary electron yield for the elastic and rediffused secondary electrons while the blue points are secondary electron yield for the true secondary electrons. Especially for molybdenum, the fitting curve has a very nice agreement with the experimental result (points in Figure 4.14). The fitting result for tungsten and molybdenum are also included in the database of ESRAY and ESPIC. The code can calculate the number, energy, and angle of secondary electrons automatically for all these three materials used in the gyrotron. Because the emitter of a cathode is usually made of porous tungsten mixed with metal oxides (i.e., BaO, CaO, and Al_2O_3) which are used to decrease the work function of the emitter, the electrons can also bombard the metal oxides. The secondary electron yield of metal oxides can be higher than pure tungsten since the work functions of metal oxides are lower. Some experimental and numerical calculation results also show such a property [Joy95, KB08, VBV+02]. However, the data resources about secondary emissions of metal oxides are limited; therefore, their secondary emissions are neglected. The emitter is set to be pure tungsten at present.

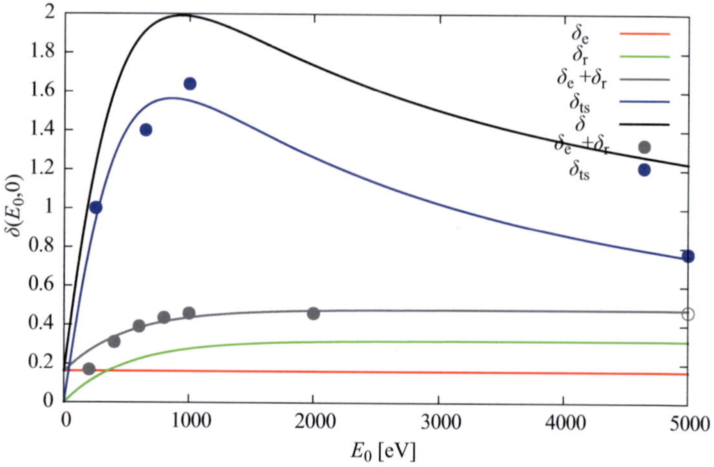

(a) The SEY of tungsten

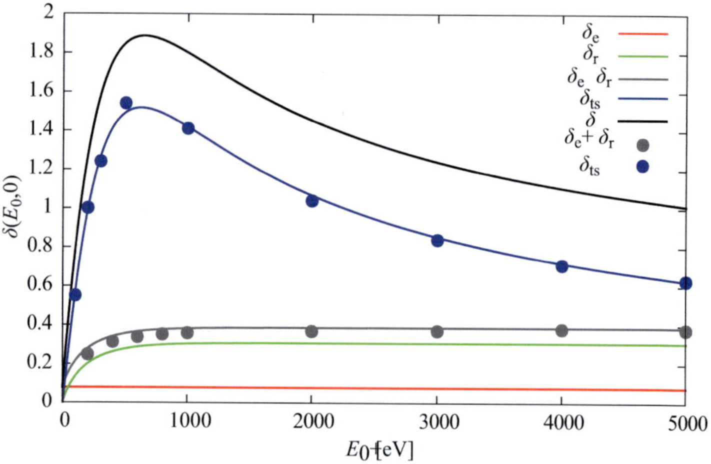

(b) The SEY of molybdenum

Figure 4.14: The SEY of tungsten and molybdenum. The dots are the experimental data from the database by Joy [Joy95]. The solid lines correspond to fitting results.

As an example, Figure 4.15 shows the distribution of electrons between the cathode and anode when back reflected electrons bombard the cathode surface. The code ESPIC is used for the PIC calculation.

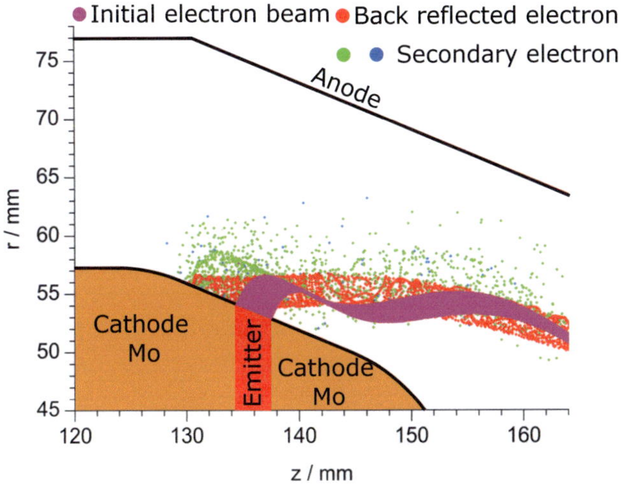

Figure 4.15: Distribution of electrons between the cathode and anode in the MIG-1 design of the EU 1 MW, 170 GHz ITER gyrotron.

As shown in Figure 4.15, secondary electrons (green and blue points) are generated after the back reflected electrons (red points) bombard the cathode surface. The secondary electrons are key factor for generation of Low Frequency Oscillations (LFOs, Chapter 5) and Electron Beam Halo (EBH, Chapter 6) in the gyrotron.

4.4 Summary

An enhanced Furman model for the generation of secondary electrons caused by the bombardment of back reflected electrons on the cathode surface is derived. In this model, the secondary emission yield and the energy spectrum is given by the Furman model while the angular distribution of elastic and rediffused electrons are modified with higher accuracy. To achieve this, the angular distribution of secondary electrons are investigated using the Monte Carlo code CASINO. The numerical investigation shows that the angular distribution of the elastic and rediffused electrons is only related to the incidence angle θ_0 of the initial electron beam. In addition, it also shows that the larger the incidence angle, more elastic and rediffused electrons have a high θ and low φ angle. This behaviour is included in the model using a polynomial surface fit up to 9×9 degrees. The calculation algorithm and calculation result for the secondary emission model are presented. The calculation result shows that copper creates a larger number of low energy secondary electrons (0-20 eV) and a smaller number of high energy secondary electrons (approximately 300 eV) compared to stainless steel. Tungsten and molybdenum are used in the cathode and their secondary electron emission parameters are deduced by fitting the existing experimental data. The enhanced secondary emission model is included in the codes ESRAY and ESPIC. The new feature of ESPIC makes it possible to investigate the deteriorating phenomena in Chapter 5 and Chapter 6.

5 Trapped electrons and low frequency oscillations in magnetron injection gun

This chapter provides the investigation on trapped electrons and Low Frequency Oscillations (LFOs) in gyrotron Magnetron Injection Guns (MIGs) based on the surface roughness and secondary electron emission models developed in Chapter 2 and Chapter 4. After a review on the mechanism of trapped electrons and LFO in MIG, the LFOs in the short-pulse prototype EU 2 MW, 170 GHz coaxial cavity gyrotron are investigated. The beam-wave instability between the electron beam and the TEM mode is analyzed based on the experimental results.

5.1 Trapping mechanism of electrons in magnetron injection guns

Due to the surface roughness, emission inhomogeneity and misalignment of a gyrotron MIG, some electrons will be reflected by a magnetic mirror and trapped in the MIG even for moderate mean pitch factor values ($\alpha \approx 1.3 - 1.5$) [MZI06]. Those reflected electrons will partly bombard the cathode surface and generate secondary electrons. Secondary electrons from the cathode surface other than the emitter surface will decrease the electron beam quality. Due to their initial energy and emission direction, the secondary electrons have higher possibilities to be trapped between the cathode and cavity, and therefore, cannot travel to the collector. The accumulation of trapped electrons in a gyrotron MIG could generate LFOs. Trapped electrons and LFOs are two main features, which deteriorate the electron beam quality by increasing the velocity and energy spreads, thus decreasing the stability and efficiency of the gyrotron [LPS+06b, CCM+09]. Since gyrotrons can never achieve a stationary status with these problems [BL97, Lou09], these phenomena can not be simulated by a static code that uses an iteration method to find the steady field in MIG. The quasi-static PIC code ESPIC [IZJ15] was used for the calculations in this chapter. The code ESPIC is an extension of the electrostatic code ESRAY, which can simulate slowly varying effects in the gyrotron MIG. The classical PIC approach is used for the particle handling, while for the calculation of electric fields in two dimensions, a multi-grid Poisson-solver is employed.

During the electrons gyration along the magnetic field line from cathode to cavity, they come up with an increasing magnetic field. As introduced in Section 1.4.1, the electrons will convert their axial velocity $\beta_{\|}$ to electrons' transverse velocity $\beta_{\perp}$. In this case, electrons may experience the magnetic mirror effect as is illustrated in Figure 5.1.

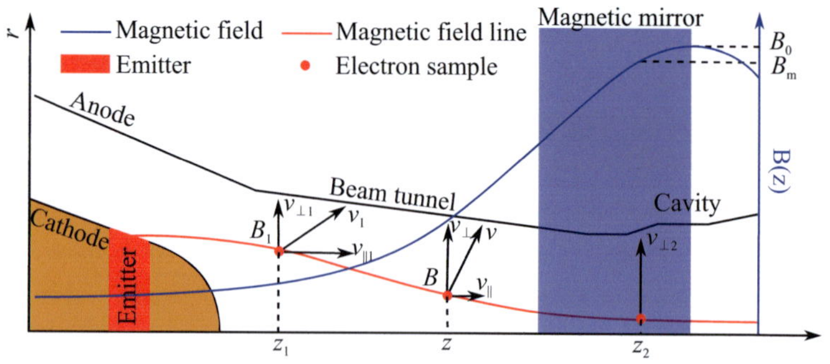

Figure 5.1: Magnetic mirror effect in a gyrotron MIG. Electrons will be reflected at z_2 when their axial velocity is zero.

According to [Siv65, Che74], if the electron gyrates in a slowly changing magnetic field (adiabatic approximation), the integral $\oint p\,dq$ is an invariant. Here, $p = mv_\perp$ is the transverse momentum and $dq = r_L d\varphi$. Then the invariant is given by

$$\oint p\,dq = 2\pi r_L m v_\perp = \frac{4\pi}{|e|} m\mu = \text{const} \tag{5.1}$$

where μ is the magnetic momentum of the electron. It is given by

$$\mu = \frac{1}{2m}\frac{p_\perp^2}{B} \quad . \tag{5.2}$$

$m = \gamma m_e$ is the relativistic mass of an electron. For the non-relativistic case $\gamma \approx 1$, the magnetic moment μ is a constant during the movement. As shown in Figure 5.1, z_1 and z are two arbitrary positions. The magnetic moment at z_1 and z follows: $\mu(z_1) = \mu(z)$. Then the relation between the transverse velocity and the magnetic field is given by

$$\frac{\frac{1}{2}mv_{\perp 1}^2}{B_1} = \frac{\frac{1}{2}mv_\perp^2(z)}{B(z)} \tag{5.3}$$

where B_1, $B(z)$ and $v_{\perp 1}$, $v_\perp(z)$ are the magnetic field and transverse velocity component at z_1 and z.

According to equation 5.4, with the increase of the magnetic field $B(z)$ along the z-axis, the kinetic energy stored in the axial velocity component is converted to the transverse velocity component. For an electron which is reflected at z_2, the axial velocity component at this position is reduced to 0. Since the kinetic energies at z_1 and z_2 are the same, the velocity at those two positions also is equal: $v_{\perp 2} = v_1$. Therefore, the magnetic moment at z_1 and z_2 is given by

$$\frac{\frac{1}{2}mv_{\perp 1}^2}{B_1} = \frac{\frac{1}{2}mv_{\perp 2}^2}{B_2} = \frac{\frac{1}{2}mv_1^2}{B_2} \quad . \tag{5.4}$$

Then, the magnetic field at z_2 is given by

$$B_2 = \frac{v_1^2}{v_{\perp 1}^2} B_1 \quad . \tag{5.5}$$

B_2 is the maximum achievable magnetic field where the electron transforms all the axial velocity to the transverse velocity. B_2 can also be named as B_m which means maximum achievable magnetic field. If the peak magnetic field at the centre of the cavity B_0 is lower than B_m, an electron can pass through the magnetic mirror. Otherwise, it will be reflected back to the cathode region. In principle, the magnetic mirror effect always exists where there is a increase of the magnetic field. In gyrotrons, the magnetic mirror effect is most obvious just in the region before the cavity centre (blue region in Figure 5.1) because there is a fast increase of the magnetic field. MIGs are designed to have no reflected electrons in the ideal case. But for the real case, there will always be some trapped electrons due to velocity spread resulting from the emitter surface roughness, emission inhomogeneity [ILM$^+$98, LPS$^+$06a], etc, which has been discussed in Chapter 2 and Chapter 3.

For a new gyrotron MIG, the amount of trapped electrons shall be very low so that they can not cause significant problems. With the increasing operation time, several factors decrease the emitter quality. Therefore, the number of trapped electrons will be increased. Firstly, during the operation of a gyrotron, barium evaporates from the emitter surface; this can increase emitter surface roughness and emission inhomogeneity. Then, due to the negative electric potential of the emitter positive ions from the ionized residual gas can bombard the emitter surface with very high energy (several tens of keV for high power gyrotrons). Bombardment of high energy ions can also increase surface roughness and emission inhomogeneity. Strong ion bombardment happens if there is a breakdown between the anode and emitter. The strong bombardment due to the breakdown can increase the surface roughness and emission inhomogeneity significantly and result in the failure of gyrotron operation. Due to the factors discussed above, the emitter quality decreases with the operation time. At some point, the increased number of trapped electrons will cause some significant problem such as LFOs [MZI06, ILM+98, CCM+09, Man09, YAN10, PLY+12, LPS+06a, LS12, VGGP12, GKM14] and Electron Beam Halos (EBHs, Chapter 6).

5.2 Space charge LFOs in the short-pulse prototype EU 2 MW, 170 GHz coaxial cavity gyrotron

During the experiment of the short-pulse prototype EU 2 MW, 170 GHz coaxial cavity gyrotron, several kinds of LFOs have been observed. Therefore, this tube has been chosen as the model to calculate the space charge LFOs.

5.2.1 Simulation setup

Figure 5.2 shows the basic sketch of the MIG for the short-pulse prototype EU 2 MW, 170 GHz coaxial cavity gyrotron. The nominal cathode-anode voltage is V_{acc} =93 kV. The beam current can be increased up to 90 A. The low-frequency oscillation in this gyrotron is numerically simulated using the code ESPIC.

The same pitch factor value as expected in the experiment $\alpha \approx 1.3 - 1.5$ is used in the calculation. The secondary electron model, which has been derived in Chapter 2, is used to take into account the effect of the bombardment by the trapped electrons. The size of the surface roughness in the simulation is assumed to be 100 μm. This "strong" surface roughness is used in order to take into account the joint influence of surface roughness, emission inhomogeneity, and misalignment. Such a "strong" disturbance has been used in the work of the Russian MIG code EPOS-V, which makes the simulated electron beam parameters agree with the experimental results [Lyg95].

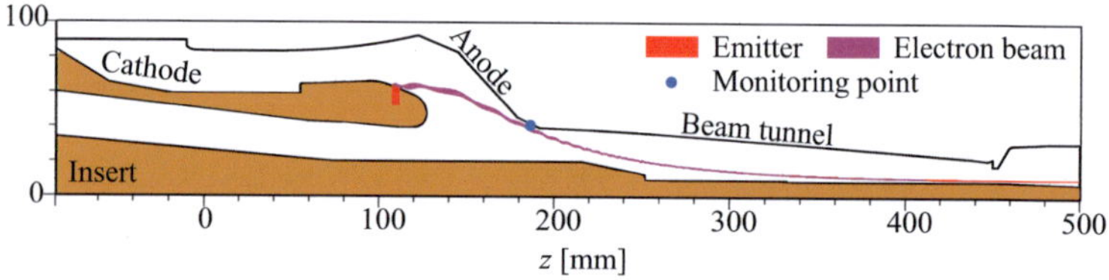

Figure 5.2: Sketch of the MIG for the short-pulse prototype EU 2 MW, 170 GHz coaxial cavity gyrotron. Beam current and electric field at the blue monitoring point is recorded during the calculation.

The Richardson-Dushman emission model with Schottky correction is used in the calculation. The variation of space charge effect due to the LFOs in the MIG will influence the electric field on the emitter surface. Using this emission model, one can take into account the influence of electric fields on the emission property so that one can get more accurate simulation results.

5.2.2 Simulation results

A sweep on beam voltage and current is made in the code ESPIC. According to the results, LFOs can be initiated if the applied voltage is between 70 to 92 kV, the beam current is 46 A, and the pitch factor $\alpha = 1.3$. Figure 5.3 depicts the time dependence of the total amount of space charge from the cathode to the cavity region. The parameters used in this calculation are: $V_{\text{acc}} = 82\,\text{keV}$, $I_{\text{b}} = 48\,\text{A}$ and $\alpha = 1.3$. According to the experimental results, the coaxial cavity gyrotron has a significant LFO under this set of parameters.

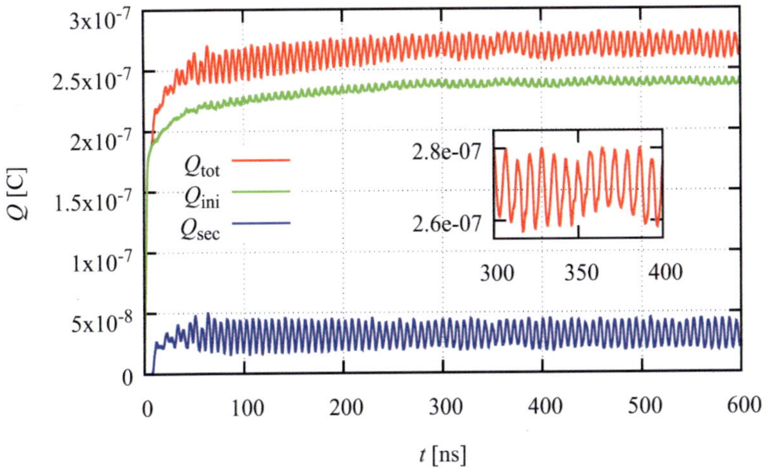

Figure 5.3: Variation of total space charge between cathode and cavity centre with time. Q_{ini} is the total charge of electrons, which were initially generated by the emitter due to thermionic emission. Q_{sec} is the total charge of secondary electrons which were generated by the bombardment of back reflected electrons on the cathode. Q_{tot} is the total charge between the cathode and the cavity centre.

The definition of the total amount of space charge Q_{tot} in Figure 5.3 is given by

$$Q_{\text{tot}} = \sum_{i=1}^{N_{\text{tot}}} q_i = \sum_{i=1}^{N_{\text{ini}}} q_{\text{ini},i} + \sum_{i=1}^{N_{\text{sec}}} q_{\text{sec},i} = Q_{\text{ini}} + Q_{\text{sec}} \tag{5.6}$$

where Q_{ini} and N_{ini} are the total charge and number of electrons, which were initially generated by the emitter due to thermionic emission. Q_{sec} and

N_{sec} are the total charge and number of secondary electrons, which were generated by the bombardment of back reflected electrons on the cathode. $q_{\text{ini},i}$ is the charge of the ith initial electron and $q_{\text{sec},i}$ is the charge of the ith secondary electron. The total number of electrons in the MIG is $N_{\text{tot}} = N_{\text{ini}} + N_{\text{sec}}$. At each time step Q_{ini}, Q_{sec} and Q_{tot} are calculated according to equation 5.6.

As is shown in Figure 5.3, the oscillation starts nearly immediately with the start of the beam. The period of the low frequency oscillation is around 7.2 ns, which is approximately the transit time of the electron from the cathode to the cavity and back to the cathode again. The amplitude of the LFOs of Q_{tot} , Q_{sec} and Q_{ini} are $3.7\,\%$, $4.6\,\%$, and $1.1\,\%$ respectively compared to the average total charge in the electron beam. The amplitude of the low frequency oscillation of the secondary electron charge Q_{sec} ($4.6\,\%$) is much higher than that for the initial electron charge Q_{ini} ($1.1\,\%$). This result indicates that secondary electron emission is a key factor in the generation of low frequency oscillations. Detailed analysis also shows that the oscillation phase of Q_{sec} and Q_{tot} are synchronous and they have a half period difference with the oscillation phase of Q_{ini}. Q_{sec} and Q_{tot} is synchronous because the oscillation of Q_{tot} mainly comes from the oscillation of Q_{sec}. The half period difference of the oscillation phase between Q_{ini} and Q_{tot} results from the Schottky effect. When Q_{tot} reaches maximum, the space charge effect of all the electrons is also most strongest during one period. The space charge effect will decrease the beam current according to the Schottky effect. Therefore, the total space charge of initial electrons Q_{ini} will be decreased. The frequency spectrum of the LFOs is plotted in Figure 5.4.

As shown in Figure 5.4, the frequency spectrum has a peak at around 140 MHz which corresponds to the period electrons emitted from the emitter, reflected by the magnetic mirror and back to the emitter again.

There is also a 276 MHz frequency, which corresponds to the second harmonic of the low frequency oscillation. The current and electric field at the blue point in Figure 5.2 is monitored during the calculation. The results are shown in Figure 5.5.

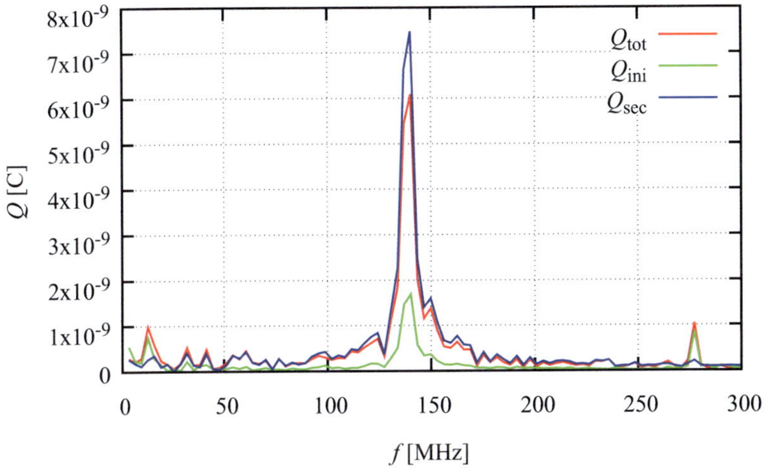

Figure 5.4: Frequency spectrum of the space charge in the gyrotron.

I is the beam current and E_r is the electric field in the radial direction at the monitor point. The beam current at this position is also oscillating. The amplitude of the low frequency oscillation of beam current is about 11.1 % compared to the average beam current. As shown in the subplot of Figure 5.5, the oscillations of the current and electric field is synchronous. Since E_r is proportional to the charge density (Gauss's flux theorem), the synchronous oscillation indicates that the oscillation of the current comes from the charge density bunching.

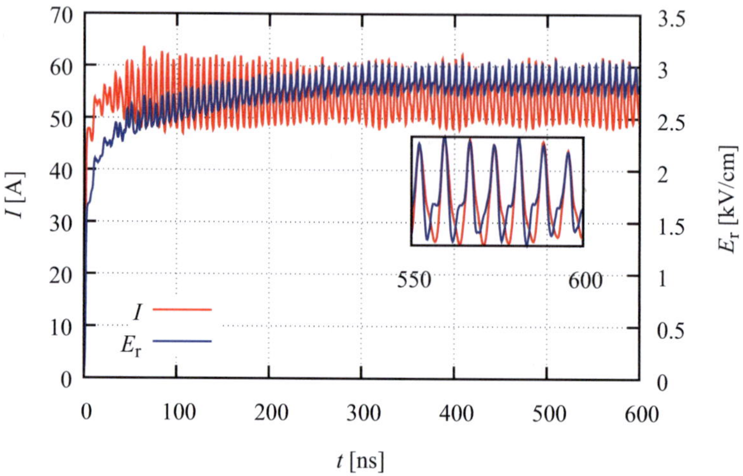

Figure 5.5: Oscillation of current (red) and electric field (blue) at the monitor point during calculation.

5.3 Eigenmode oscillation in the short-pulse prototype EU 2 MW, 170 GHz coaxial cavity gyrotron

As shown in Figure 5.2 there is a coaxial insert in the coaxial cavity gyrotron. The coaxial insert together with the gyrotron body forms a coaxial waveguide. TEM modes can be excited in the coaxial waveguide. Unlike the cut-off frequency for TE or TM modes which are determined by the structure of the gyrotron there is no cut-off frequency for the TEM modes in a coaxial waveguide. Therefore, TEM modes can resonate in the axial direction of the MIG with a very low frequency (in 100 MHz range). TEM modes can interact with the travelling electron beam also. Once the beam-wave instability condition is fulfilled, the TEM mode can gain energy from the electron beam and is excited. During the experiment,

LFOs of TEM modes can interfere with the power supply system, which results in the failure of the operation. A similar interaction between the LFOs of space charge and power supply system was observed by General Atomics [GLP+99]. In this Section, the resonance conditions for the TEM mode in the coaxial gyrotron and the beam-wave instability mechanism is analyzed.

5.3.1 Resonance conditions in the coaxial cavity gyrotron

The 2-D schematic diagram [Rze15] of the coaxial cavity gyrotron is shown in Figure 5.6. The brown colour parts are the metal parts while the blue parts are the ceramic components.

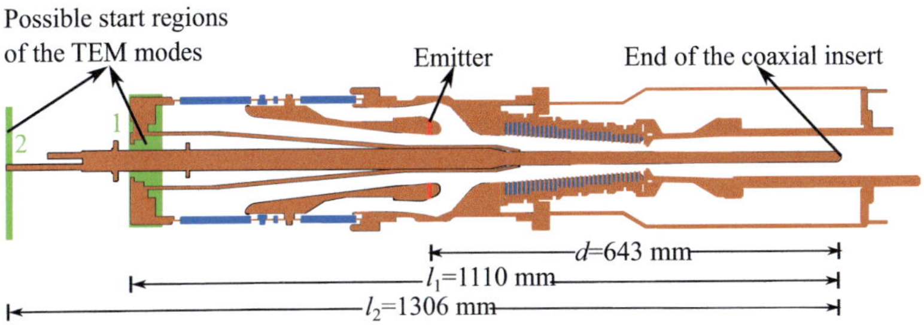

Figure 5.6: 2-D schematic diagram of the coaxial cavity gyrotron.

The first start region of the TEM mode is a closed boundary, which is shown in the first green region in Figure 5.6. The second start region of the TEM mode is an open boundary, which is shown in the second green region in Figure 5.6. Since TEM modes cannot propagate in cylindrical waveguides those modes stop at the end of the coaxial insert. Therefore, the coaxial gyrotron can be simplified into two different coaxial resonators, as shown in Figure 5.7.

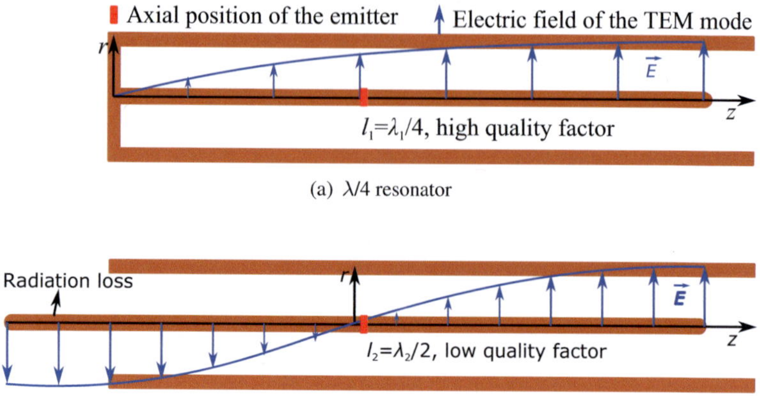

(a) $\lambda/4$ resonator

(b) $\lambda/2$ resonator

Figure 5.7: Resonator model (the red mark stands for the position of the emitter).

For the first type of resonator (see Figure 5.7(a)), which has a closed boundary on the left side and an open boundary on the right side, the relation between resonant wavelength and length of the coaxial insert are given by [LYG04]

$$\lambda_1 = \frac{4l_1}{2p-1} \qquad (p = 1, 2, 3, \dots) \qquad . \tag{5.7}$$

Here, l_1 is the length of the coaxial insert for the first kind resonator. λ_1 is its resonant wavelength.

If $p = 1$, $l_1 = \lambda_1/4$: The coaxial gyrotron behaves like a $\lambda/4$ resonator with electric wall boundary condition. As plotted in Figure 5.6, the length $l_1 = 1.11$ m. The resonant wavelength and the frequency for the first kind resonator are: λ_1 =4.44 m and f_1 =67.5 MHz.

For the second type of resonator (see Figure 5.7(b)), which has two open boundaries the relation between resonant wavelength and length of the coaxial insert are given by [Chu15]

$$\lambda_2 = \frac{2l_2}{p} \qquad (p = 1, 2, 3, \dots) \qquad . \tag{5.8}$$

Here, l_2 is the length of the coaxial insert for the second kind resonator and λ_2 is its resonant wavelength.

If $p = 1$, $l_2 = \lambda_2/2$: The coaxial gyrotron behaves like a $\lambda/2$ resonator with electric wall and open boundary condition. According to the data which are given in Figure 5.6 with $l_2 = 1.306\,\text{m}$, the resonant wavelength and the frequency for the second kind resonator are: λ_2 =2.612 m and f_2 =114.8 MHz.

For both the $\lambda/4$ and the $\lambda/2$ resonators, there is no radiation loss at the right part of the resonator due to the extended outer radius. But for the left part of the $\lambda/2$ resonator, the insert part is longer than the outer part. The exposed part of the insert, which has no metallic surroundings, leads to a coupling of the resonator and surrounding environment. The coupling results in radiation loss from this part. Therefore, the $\lambda/4$ resonator has a higher quality factor Q than the $\lambda/2$ resonator.

5.3.2 Instability of the beam-wave interaction

With the periodical influence of the electric field of the TEM mode and even the oscillating space charge, a beam-wave instability can be excited. The exchange of energy between the electron beam and the $\lambda/4$ TEM mode is shown in Figure 5.8.

$\vec{E}_0$ is the electric field which is applied between the cathode and the anode. It is determined only by the cathode-anode voltage and structure of MIG, it has a constant value and direction. $\vec{E}_{\text{TEM}}$ is the electric field of the TEM mode which has a periodical changing value and direction. Therefore, the total electric field is

$$\vec{E}_{tot} = \vec{E}_0 + \vec{E}_{\text{TEM}} \qquad . \tag{5.9}$$

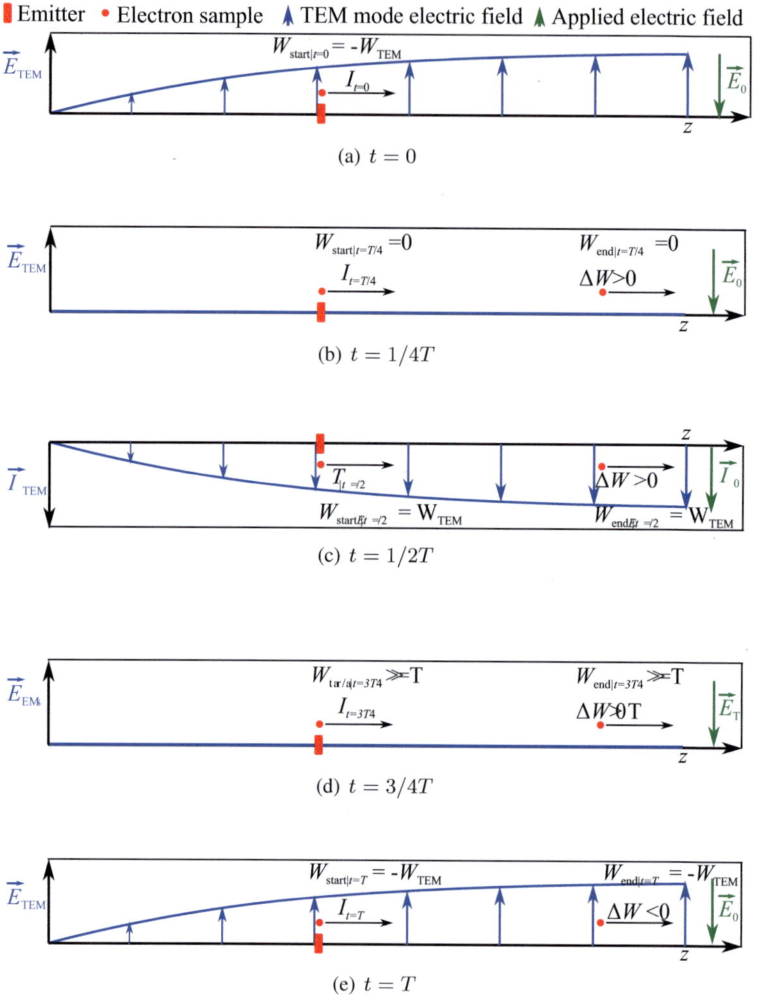

Figure 5.8: Energy exchange process between the electron beam and the $\lambda/4$ TEM mode for different time steps during one oscillation period.

The current of the electron beam is determined by the value of the total electric field $|\overrightarrow{E}_{tot}|$. Since $\overrightarrow{E}_0$ has a constant value and direction, the beam current will change with the variation of $\overrightarrow{E}_{\text{TEM}}$. When $\overrightarrow{E}_0$ and $\overrightarrow{E}_{\text{TEM}}$ have the same direction, there will be a higher electron beam current than when they have opposite directions. In case of $\overrightarrow{E}_{\text{TEM}} = 0$, the beam current is I_0.

The beam-wave interaction processes in Figure 5.8 are:

(a) $t = 0$: The direction of $\overrightarrow{E}_{\text{TEM}}$ is opposite to $\overrightarrow{E}_0$. The beam current is $I_{t=0} < I_0$. The electrons emitted at this time have additional negative potential energy $-W_{\text{TEM}}$.

(b) $t = 1/4T$: The amplitude of $\overrightarrow{E}_{\text{TEM}}$ is 0. The beam current is $I_{t=1/4T} = I_0$. The electrons emitted at this time have no additional potential energy. The electrons emitted at $t = 0$ also have no additional potential energy at this time, which means that they gain energy from the RF wave.

(c) $t = 1/2T$: The direction of $\overrightarrow{E}_{\text{TEM}}$ is the same as for $\overrightarrow{E}_0$. The beam current is $I_{t=1/2T} > I_0$. The electrons emitted at this time have additional positive potential energy W_{TEM}. The electrons emitted at $t = 1/4T$ also have positive additional potential energy at this time which means that they get energy from the RF wave.

(d) $t = 3/4T$: The amplitude of $\overrightarrow{E}_{\text{TEM}}$ is 0. The beam current is $I_{t=3/4T} = I_0$. The electrons emitted at this time have no additional potential energy. The electrons emitted at $\frac{1}{2}T$ also have no additional potential energy at this time which means that they lose energy to the RF wave.

(e) $t = T$: The direction of $\overrightarrow{E}_{\text{TEM}}$ is opposite to $\overrightarrow{E}_0$. The electrons emitted at $t = 3/4T$ also have negative additional potential energy at this time which means that they give energy to the RF wave.

Electrons emitted at $t = 0$ and $t = 1/4T$ with the beam current $I_{t=0}$ and $I_{t=1/4T}$ will get energy from the RF wave. Electrons emitted at $t = 1/2T$ and $t = 3/4T$ with the beam current $I_{1/2T}$ and $I_{t=3/4T}$ will lose energy to the RF wave. Due to the Schottky effect, the relations between the current at different time are $I_{t=0} < I_{t=1/4T} = I_{t=3/4T} < I_{t=1/2T}$. As a result, the average energy the electron beam gives to the RF wave is positive. Due to the beam-wave instability mechanism illustrated above one possible solution to suppress it is to move the emitter position to the left of the point where $|\overrightarrow{E}_{\text{TEM}}| = 0$. In this case the TEM mode will lose energy to the electron beam in one period, therefore, the TEM mode can not be excited.

The same beam-wave instability mechanism exists for the TEM mode in the $\lambda/2$ resonator. But for the $\lambda/2$ resonator, the emitter position is closer to the point where $|\overrightarrow{E}_{\text{TEM}}| = 0$ compared to the $\lambda/4$ resonator, as shown in Figure 5.7. According to the analysis of the beam-wave instability mechanism, the electric field $\overrightarrow{E}_{\text{TEM}}$ is the key factor for the energy exchange. If the $|\overrightarrow{E}_{\text{TEM}}|$ at the emitter position is lower, the coupling between electron beam and RF wave is lower. Therefore, the coupling between electron beam and RF wave in the $\lambda/2$ resonator is lower than in the $\lambda/4$ resonator. According to the analysis in Section 5.3.1 and Section 5.3.2, the information for the $\lambda/4$ resonator and $\lambda/2$ resonator are shown in Table 5.1.

5.3.3 Experimental Results

An antenna is used to measure the electromagnetic signal around the gyrotron during the operation. Figure 5.9 shows the typical measured waveform and frequency spectrum of the TEM modes for the $\lambda/4$ and $\lambda/2$ resonator.

Table 5.1: Parameters for the $\lambda/4$ and $\lambda/2$ resonator.

Mode	**Frequency [MHz]**	**Quality factor**	**Beam-wave coupling**
$\lambda/4$	67.5	Higher	Higher
$\lambda/2$	114.8	Lower	Lower

The red line in Figure 5.9 is the measured signal and frequency spectrum when both TEM modes are excited. The blue line is the measured signal and frequency spectrum when the TEM mode is excited only in the $\lambda/2$ resonator. The measured frequencies are 64.9 MHz and 114.8 MHz, which is quite near from the analyzed frequency shown in Table 5.1. Under high power gyrotron operation, the amplitude of the 64.9 MHz TEM mode is much larger than that of the 114.8 MHz TEM mode. This is due to two facts which are analyzed in Section 5.3.2. Firstly, the radiation loss of the $\lambda/2$ TEM mode (114.8 MHz) is higher than that of the $\lambda/4$ TEM mode (64.9 MHz). Secondly, the beam-wave coupling coefficient of the the $\lambda/2$ TEM mode (114.8 MHz) is much lower than that of the $\lambda/4$ TEM mode (64.9 MHz). During the experiment, the $\lambda/4$ TEM mode (64.9 MHz) can interfere with the power supply system which results in strong oscillations in the system and prevents the system to operate at higher accelerating voltage. This phenomenon has been suppressed by adding an absorber near the oil tank around the magnetron injection gun to reduce the quality factor of the resonances inside the gyrotron circuit [RPK+10]. The absorbing structure consists of a metallic cylinder filled with the damping material Eccosorb.

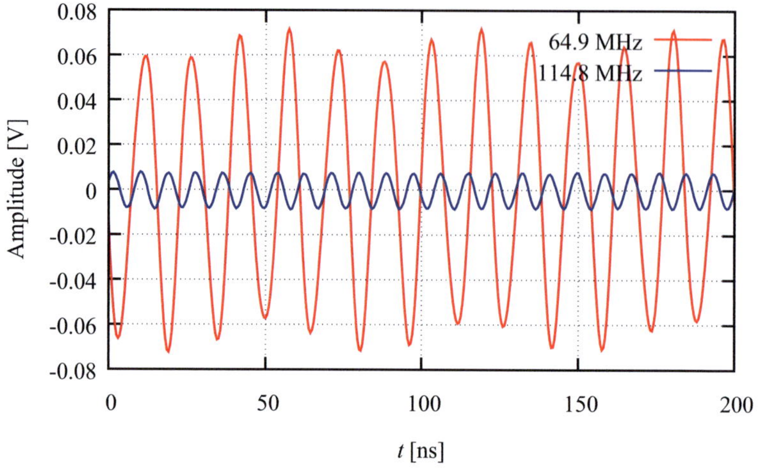

(a) Measured oscillation signals

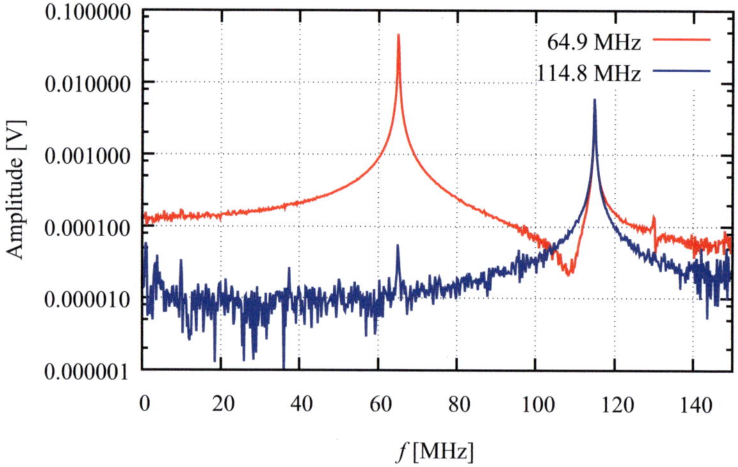

(b) Frequency spectrum of the two measured signals

Figure 5.9: Measured signal and frequency spectrum of the two TEM modes.

It is also found during the experiment that the 114.8 MHz signal exists in a much wider operating range compared to the 64.9 MHz signal. This is due to the fact that the LFOs of the trapped electrons is between 110 MHz and 140 MHz, which means that the electron beam has an additional excitation source for the 114.8 MHz TEM mode. Due to the additional excitation from the LFOs of trapped electrons, the 114.8 MHz TEM mode is excited in such a wide operating range.

5.4 Summary

The trapping mechanism of electrons in the magnetron injection gun is presented in detail. The space charge LFO in the short-pulse prototype EU 2 MW, 170 GHz coaxial cavity gyrotron caused by back reflected electrons is investigated using the code ESPIC. To investigate space charge LFOs, the surface roughness model (Chapter 2) and the secondary emission model (Chapter 4) are used in the calculation. A "strong" surface roughness is used take into account all the influence of surface roughness, emission inhomogeneity, and misalignment. The numerical investigation shows that space charge LFOs can be generated by the trapped electrons. The period of the low frequency oscillation is approximately the transit time of the electron from the cathode to the cavity and back to the cathode again.

The beam wave instability mechanism between the electron beam and the TEM modes are illustrated. The net energy exchange between electron beam and TEM modes results from the Schottky effect. Theoretical analysis indicates that the $\lambda/4$ TEM mode has higher quality factor and beam-wave coupling coefficient compared to the the $\lambda/2$ TEM mode. The measured results of the TEM modes have a good agreement with the theoretical analysis. The experimental results also indicate that space charge LFOs of the trapped electrons is an additional excitation source for the TEM mode. One possible solution to suppress the TEM mode is to move the emitter position to the left of the point where $|\overrightarrow{E}_{\mathrm{TEM}}| = 0$. In this case the TEM mode will lose energy to the electron beam in one period, as a consequence, the TEM mode cannot be excited.

6 Electron beam halo in gyrotron MIGs

In this chapter, the generation mechanism of Electron Beam Halos (EBHs) has been studied. EBHs generation process in two different gyrotron Magnetron Injection Guns (MIGs) are investigated. The simulation results have been correlated with experimental results. To suppress the EBHs generation in gyrotron MIGs, three solutions are proposed at the end of this chapter.

6.1 Electron beam halo generation mechanism

It has been experimentally shown that even for average pitch factors less than one, a population of adiabatically trapped electrons was observed [LSK99]. The velocity spread could be much higher than the expected one due to many factors such as emitter surface roughness (Chapter 2), emission inhomogeneity (Chapter 3), manufacturing tolerances and misalignments. Therefore, a small population of the beam electrons could have initial conditions which cause them to be back reflected and adiabatically trapped by the magnetic mirror. Back reflected electrons from the main electron beam will bombard the cathode surface and generate secondary electrons. As discussed in Chapter 5, parts of these secondary electrons can also be trapped in the gyrotron. Accumulation of trapped electrons in the gyrotron MIG can result in some deteriorating phenomena, such as EBH. Electron traces at many positions on the surface of the gyrotron components were observed during the inspection of the first industrial prototype of the EU 2 MW, 170 GHz coaxial cavity gyrotron after the end of the experiments. A correlation of the positions of most of the damaged spots with the EBH has been found [PHG+09, PPZ+16].

The EBH is generated by the concentration of trapped electrons around the main electron beam. In the design of MIGs, the pitch factor α at the cavity centre of the beam electrons is kept to be at a level lower than 1.5. However, the pitch factor α of the electrons emitted from neighbouring parts of the emitter, such as the cathode nose and the rear part (or prolongator), can be much higher, as shown in Figure 6.1. Figure 6.1(b) shows the "cold" pitch factor α at the cavity versus the axial position of the emitting point on the cathode surface (blue line in Figure 6.1(a)). Here, "cold" means the space charge of the electrons is not considered.

As introduced in Section 5.1, the higher the α value, the lower is the longitudinal velocity component. For the extreme case, the longitudinal velocity component of the electrons can be decreased to zero before they arrive at the point where the magnetic field has its maximum. Then, this electron will be back reflected to the cathode region. The region from where these back reflected electrons are emitted is defined as "back reflected region", as shown in Figure 6.1. For the ideal case, no electrons are emitted from this region. However, in reality, secondary electrons generated by the bombardment of back reflected electrons from the main beam can extend the bombarded area on the cathode surface to the back reflected region. Electrons that are emitted from the back reflected region on the cathode surface will be trapped between the cathode and magnetic mirror. The accumulation of trapped electrons in this region can generate an EBH around the main electron beam.

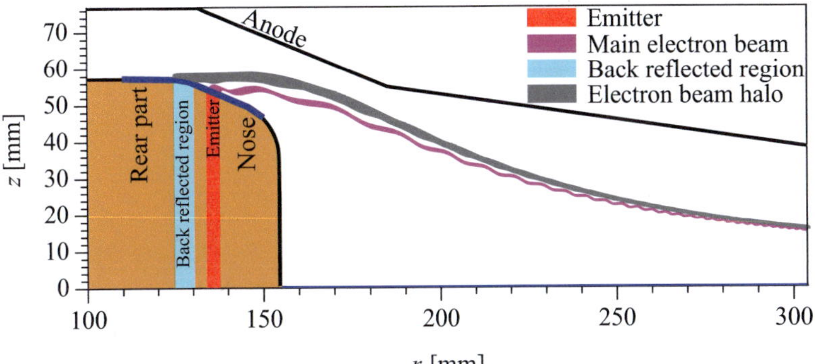

(a) Electron beams above the cathode surface.

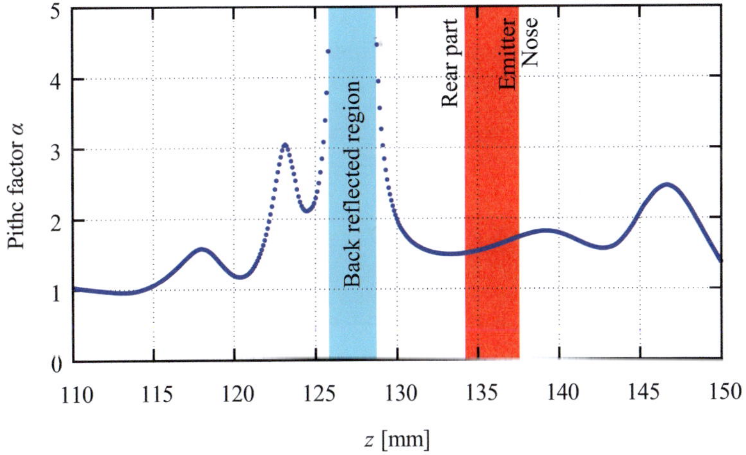

(b) The "cold" pitch factor α at the cavity centre.

Figure 6.1: Sketch of (a) electron beams and the "cold" pitch factor α at the cavity versus the axial position of the emitting point of the MIG-1 design of the EU 1 MW, 170 GHz ITER gyrotron.

Therefore, the EBH is defined as an electron beam that is generated by the trapped electrons emitted from the back reflected region of the cathode in the gyrotron [PHG+09, PPZ+16], as shown in Figure 6.1(a). The magenta coloured beam is the main electron beam and the gray beam is the EBH generated from the back reflected region of the cathode.

6.1.1 Successive and cascading generation of secondary electrons along the cathode surface

Back reflected electrons from the main electron beam can only bombard the area close to the the emitter. As analysed above, the extension of the bombarded area on the cathode surface from the emitter to the back reflected region is the key factor which initializes the generation of an EBH. Successive and cascading generation of secondary electrons along the cathode surface is the reason for the extension of the bombarded area, as shown in Figure 6.2.

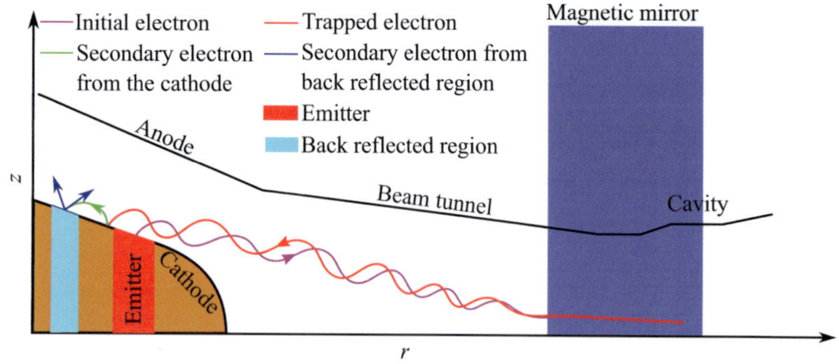

Figure 6.2: Successive and cascading generation of secondary electrons along the cathode surface.

As introduced in Section 5.1, in gyrotrons, the magnetic mirror effect is most obvious before the cavity centre (blue region in Figure 6.2) and the electrons are back reflected by the magnetic mirror effect in this region. When electrons from the main electron beam (magenta line) are reflected (red line) to the cathode surface due to the high α value caused by surface roughness or emission inhomogeneity most of the back reflected electrons will bombard the cathode surface which is on the rear part (left part) of the emitter [MZI06]. The generated secondary electrons from the bombarded area have velocities pointing to all the directions above the cathode surface. Instead of going into the beam tunnel and cavity, parts of the secondary electrons which have initial velocity pointing to the $-z$ direction (green line) can bombard the cathode surface again. The bombardment of secondary electrons on the cathode surface will also generate secondary electrons. Consequently, the successive and cascading generation of secondary electrons can be extended to the back reflected region. At the end, the secondary electrons will accumulate in this region until the space charge effect of the trapped electron is strong enough to decrease α to a sufficiently low value. Under this condition, the rate of electrons generated in the back reflected region is equal to the rate of electrons that leave this region and the system reaches a stable situation.

Verification of the successive and cascading generation of secondary electrons along the cathode surface

To simulate the successive and cascading generation process of secondary electrons, the code ESPIC with surface roughness model and secondary electron emission model developed in Chapter 2 and Chapter 4 is used. As is shown in Figure 6.1, there is a back reflected region on the cathode of the MIG-1 design for the EU 1 MW, 170 GHz ITER gyrotron. The average α value at the cavity centre of the main electron beam is 1.5 and the beam current is 40.0 A. To generate back reflected electrons, the surface roughness microstructure size of the emitter is set to be 50 μm. 40 macro particles are emitted from the emitter at every time step, so as to demonstrate the successive and cascading generation of secondary electrons at an acceptable level of simulation time. It was verified that, in this case, ESPIC can get the same beam parameters as ESRAY. When the back reflected electrons are about to bombard the cathode surface they are separated into 100 macro particles. With this approach one can simulate the case when the cathode is bombarded by a large amount of back reflected electrons without increasing the number of electrons in the main beam.

Figure 6.3 is the verification of the successive and cascading generation process of secondary electrons. Figure 6.3(a) shows that parts of the electrons are reflected to the cathode region by the magnetic mirror at 8.2 ns, which is the period for the electrons emitted from the emitter — to propagate to the cavity — and get back reflected to the cathode surface by the magnetic mirror. When the trapped electrons bombard the cathode surface secondary electrons are generated. Parts of the secondary electrons can bombard the cathode surface again. Figure 6.3(b) shows the distribution of secondary electrons above the cathode surface. The colour of the electrons stands for the generation of the secondary electrons. The blue particles are the first generation of the secondary electrons while the red particles are the sixth generation of the secondary electrons. It is shown that the successive and cascading generation of secondary electrons is extended to the back reflected region in 1.0 ns. Although there are lots of secondary electrons in Figure 6.3(b), it shall be pointed out that their charge is much smaller compared to the charge of the primary electrons in Figure 6.3(a). After the secondary electrons are emitted from the back reflected region they will be trapped in the gyrotron MIG. As expected, accumulation of these secondary electrons can generate an EBH.

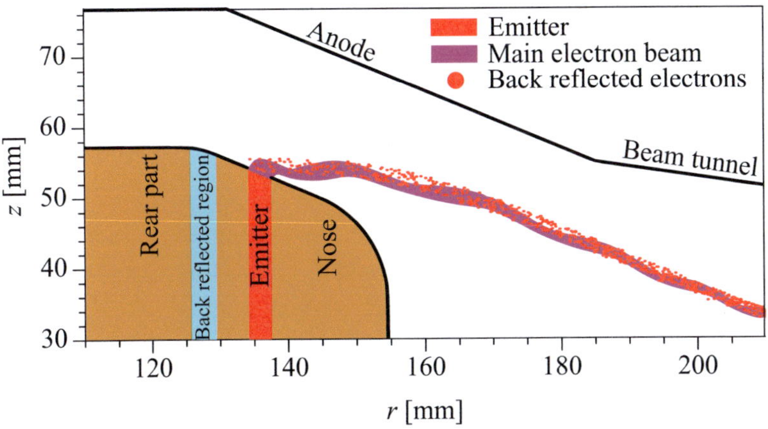

(a) Electrons above the cathode surface at 8.2 ns

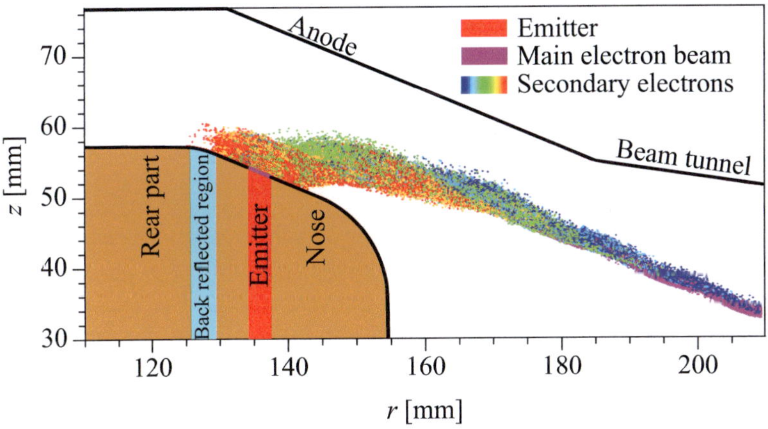

(b) Electrons above the cathode surface at 9.2 ns

Figure 6.3: Verification of the successive and cascading generation of secondary electrons on the cathode surface of the MIG-1 design of the EU 1 MW, 170 GHz ITER gyrotron.

6.1.2 Electron beam halo in the MIG-1 design of the EU 1 MW, 170 GHz ITER gyrotron

The generation process of the EBH initiated by the back reflected electrons from the main beam in the MIG-1 design of the EU 1 MW, 170 GHz ITER gyrotron is shown in Figure 6.4.

The electron beam parameters are the same as given in Section 6.1.1. However, 100 macro particles are emitted at every time step and these electrons, which bombard the cathode surface, are not separated. This new setup is suitable for long time calculations, which can demonstrate the generation process of the EBH.

The primary electrons (magenta) are emitted by the emitter while the secondary electron beam is plotted in red colour. At the very beginning, only a primary electron beam exists in the gyrotron MIG. After the bombardment of back reflected electrons from the magnetic mirror secondary electrons begin to climb along the cathode surface. At 0.2 μs, a few secondary electrons climb up to the back reflected region and then between 0.2 μs and 1.0 μs, more and more secondary electrons emitted from this region begin to accumulate in the gyrotron. Figure 6.5 is the total space charge between the cathode and the cavity centre.

As is shown in Figure 6.5, the space charge of secondary electrons Q_{sec} is nearly constant before 0.2 μs. After the secondary electrons climb up to the back reflected region at 0.2 μs, Q_{sec} has a fast increase. This is due to the fact that secondary electrons from the back reflected region will be trapped in the gyrotron. At 1.0 μs, the total space charge of the secondary electrons increases to around $50\,\%$ of the main beam space charge Q_{ini}. The strong space charge effect of electrons at the back reflected region will reduce the electric field strength on the cathode surface. As a consequence, the average pitch factor at the cavity centre of the secondary electrons generated from the back reflected region will be decreased. The decreased α value will decrease the generation rate of secondary electrons between the cathode and magnetic mirror. At 1.5 µs, the generation rate is equal to the rate of secondary electrons that leave this region. The space charge of the secondary electron stops increasing and reaches a quasi-stable state.

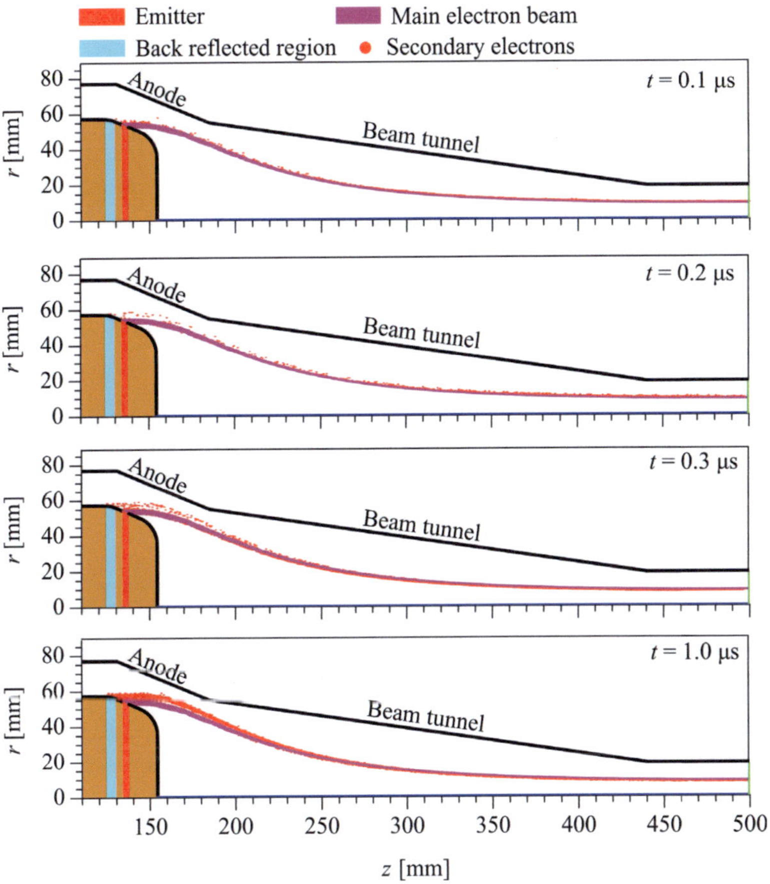

Figure 6.4: Time variation of electron beam in the MIG-1 design of the EU 1 MW, 170 GHz ITER gyrotron.

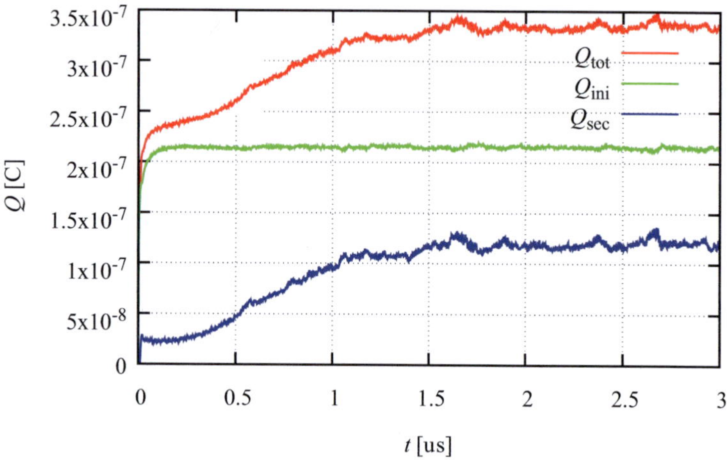

Figure 6.5: Variation of the total space charge in the MIG-1 design of the EU 1 MW, 170 GHz ITER gyrotron.

6.1.3 Electron beam halo in the first industrial prototype of the EU 2 MW, 170 GHz coaxial cavity gyrotron

During the inspection of the first industrial prototype of the EU 2 MW, 170 GHz coaxial cavity gyrotron after the end of the experiment, electron traces at many positions on the surface of the gyrotron components were observed. A correlation of the positions of most of the damaged spots with the EBH has been proposed [PHG$^+$09]. The cathode-anode geometry of the first industrial prototype of the EU 2 MW, 170 GHz coaxial cavity gyrotron is illustrated in Figure 6.6(a). The "cold" α along the cathode surface is shown in Figure 6.6(b).

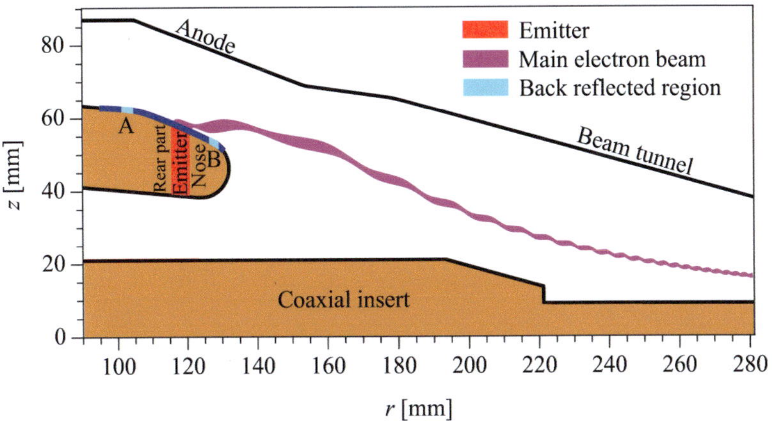

(a) Cathode-anode geometry

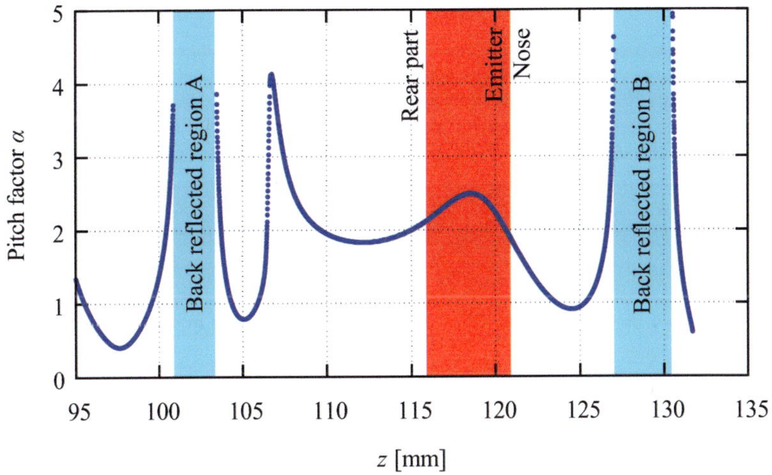

(b) The "cold" pitch factor α at the cavity centre

Figure 6.6: Sketch of (a) cathode-anode geometry and (b) the "cold" pitch factor α at the cavity versus the axial position of the emitting point of the first industrial prototype EU 2 MW, 170 GHz coaxial cavity gyrotron.

Regions A and B are two back reflected regions beside the emitter. According to [PHG+09], the damage at the end of the launcher of the quasi-optical output coupler is correlated with the EBH from the left part of the emitter (region A in Figure 6.6(a)), while the damage on the coaxial insert and many electron traces on the cathode nose could be correlated with the EBH from the right part of the emitter (region B in Figure 6.6(a)).

In order to verify the generation of these two EBHs in the coaxial gyrotron, PIC calculations similar to the setup for the EU 1 MW, 170 GHz ITER gyrotron were made using the code ESPIC. The evolution of the EBH in this coaxial gyrotron is shown in Figure 6.7.

To generate back reflected electrons from the main electron beam, which initiate the EBH, a strong surface roughness of the order of 50 μm is used in the calculation. In the calculation shown in Figure 6.7, the applied voltage between cathode and anode is 93 kV, the beam current is 46 A, and the Richardson-Dushman emission model with Schottky modification is used. In Figure 6.7, it is shown that a few trapped electrons begin to appear between the emitter region and the left back reflected region at 1.0 μs. At 1.5 μs, some electrons already climbed to the back reflected region A and there is a fast increase of electrons between 1.5 μs and 2.0 μs. The EBH in region A is generated at 2.0 μs. Between 2.0 μs and 4.0 μs, more electrons are trapped in the gyrotron. Detailed information about the formation of EBH and the increase of the space charge in the gyrotron are shown in Figures 6.8 and 6.9.

As shown in Figure 6.8 and Figure 6.9, after the electrons reach the left back reflected region, there is an exponential growth between 1.5 μs and 2.0 μs. At this period, the electrons accumulate very quickly in the back reflected region, which results in a strong space charge effect that decreases the electric field above the emitter surface significantly. Therefore, the main beam current and average pitch factor α will decrease, as a result, the space charge from the emitter Q_{ini} will decrease and the number of back reflected electrons from the main beam will also decrease.

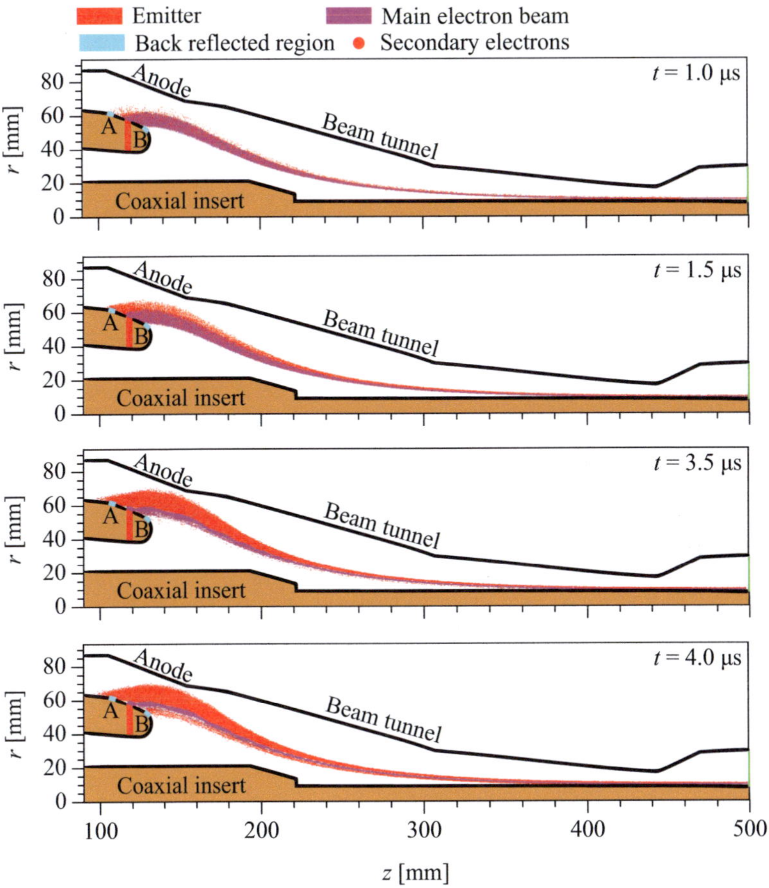

Figure 6.7: Time variation of EBH in the first industrial prototype EU 2 MW, 170 GHz coaxial cavity in gyrotron.

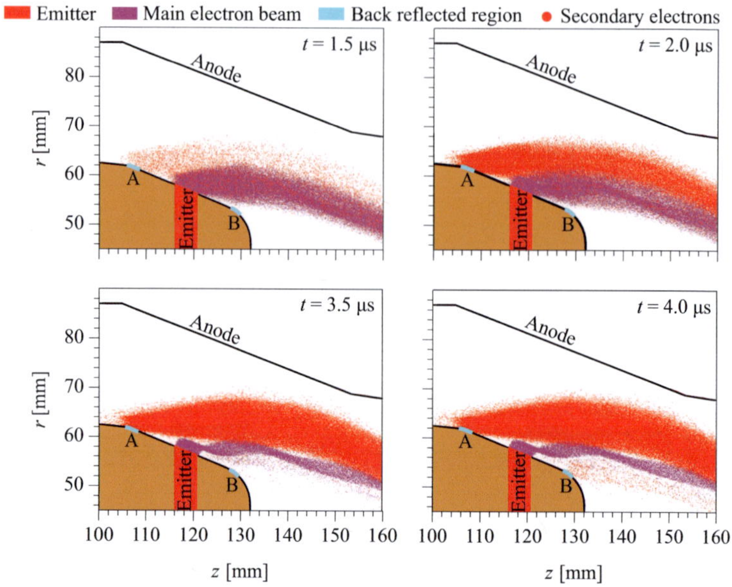

Figure 6.8: Time variation of secondary electrons above the cathode surface.

From the comparison of the electrons distribution at 1.5 μs and 3.5 μs it is clear that after the formation of the EBH in region A (3.5 μs) the number of trapped electrons from the main beam is significantly decreased. After 3.5 μs, the EBH in region B begins to appear and is generated between 3.5 μs and 4.0 μs. The generation of the inner EBH in region B is also caused by the secondary electrons emitted from the cathode surface. Because most of the back reflected electrons bombard the rear part of the emitter (between the emitter and region A) the generation of EBH in region B is slower than in region A. During the formation of the EBH in region B (3.5-4.0 μs) the total space charge increases slower than that of region A. As shown in Figure 6.8 the number of electrons of the EBH in region B is significantly smaller than that of region A at 4.0 μs. This is caused by the fact that the strong space charge effect from the EBH in region A decreases the pitch factor α of electrons emitted from region B. As a result, the number of trapped electrons is also decreased in this region. These two EBHs generated in region A and region B confirm that EBHs can be generated in both the back reflected region of the first industrial prototype EU 2 MW, 170 GHz coaxial cavity gyrotron. Electrons escaped from the EBHs can bombard the gyrotron components and result in damages. Damages spots in this gyrotron has been observed during the inspection after the end of the experiment [PHG⁺09]. The damage at the end of the launcher of the quasi-optical output coupler is correlated with the EBH generated from the left part of the emitter (region A), while the damage on the coaxial insert and many electron traces on the cathode nose is correlated with the EBH generated from the right part of the emitter (region B).

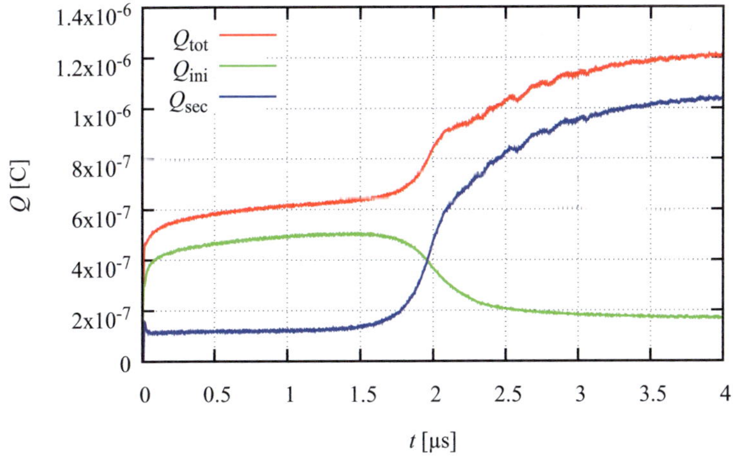

Figure 6.9: Development of the total space charge in the gyrotron for the case of "large" radius secondary electron beam.

6.2 Electron beam halo initiated by thermionic emission

It is pointed out in the above study that electrons emitted from the back reflected region can generate an EBH in the gyrotron MIG. Besides the back reflected electrons from the main electron beam, electrons can also emit from the cathode surface due to some other mechanisms, such as thermionic emission. Thermionic emission will start if the cathode surface is polluted by barium. These thermionic emitted electrons will also be trapped and generate secondary electrons. Therefore, an EBH could also be produced due to this effect. To study the possibility of EBH generated by thermionic emission, its emission mechanism of electrons from the back reflected region is investigated first. The parameters for three kinds of dispenser emitters [Fal83], which are used in high power gyrotrons, are shown in Table 6.1.

Table 6.1: Typical emitters used in high power gyrotrons.

Type	**Description**	**Work function**	**Typical working temperature**
B	Porous tungsten impregnated with 5:3:2 $BaO \cdot CaO \cdot Al_2O_3$	2.1 eV	1300 K
S	Porous tungsten impregnated with 4:1:1 $BaO \cdot CaO \cdot Al_2O_3$	2.1 eV	1300 K
M	B type coated with osmium/ruthenium	1.8 eV	1200 K

The typical working temperature of B-type, S-type, and M-type emitters are given in the case of a current density of 2.5 A/cm^2. According to [Cro81, NGM$^+$56], barium can evaporate from the emitter surface due to the high temperature. Then, it can adhere to the surface of other parts of the cathode where the temperature is lower. The other parts of the cathode are made of molybdenum. Typical work functions of molybdenum and barium are 4.35-4.9 eV and 2.52-2.7 eV, respectively [JC71]. After adherence of barium on the molybdenum surface, the work function of molybdenum will decrease. To simulate the strongest influence of the thermionic emission from the back reflected region, the lowest possible work function value (B-type emitter) is taken for the back reflected region. The MIG-1 design of the EU 1 MW, 170 GHz ITER gyrotron [PIS$^+$08] which has a back reflected region on the cathode surface is used. Figure 6.10 shows the EBH (blue) generated in the back reflected region compared to the main electron beam (magenta).

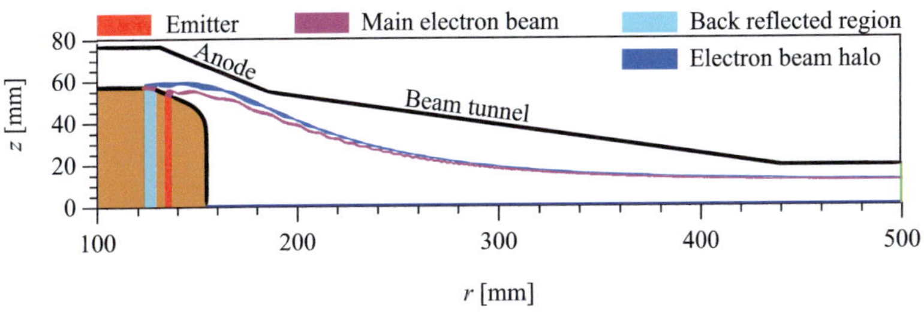

Figure 6.10: Sketch of electron beam halo initiated by thermionic emission.

The surface roughness of the emitter is neglected so that there are no back reflected electrons from the main beam which will influence the generation of the EBH. The typical temperature of the cathode is 900 K [Pag15]. In this case, the thermionic emission current density on the back reflected region is 6.36×10^{-4} A/cm^2. The total current from the back reflected region is 1.58 mA. The simulation result of the total space charge which is emitted from the back reflected region and trapped between the cathode and magnetic mirror is shown in Figure 6.11.

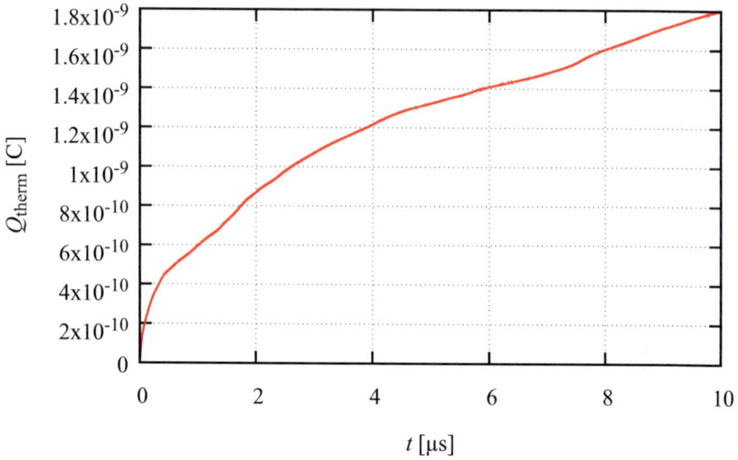

Figure 6.11: Total space charge between the cavity and the cathode of electrons emitted from the back reflected region.

It is shown that the total space charge from the thermionic emission in the back reflected region Q_{therm} is still less than 1.8×10^{-9} C at 10.0 µs. However, the total space charge of the EBH initiated by back reflected electrons from the main beam (Q_{sec} shown in Figure 6.5) at 3.0 µs is around 1.2×10^{-7} C. It shall be pointed out that the most "aggressive" parameters are used in our calculation which means the lowest work function value and the highest temperature. In the normal case the thermionic emission current shall be significantly lower. Even in this case, it is shown that back reflected electrons from the main electron beam have a faster effect on the generation of EBH compared to thermionic emission from the back reflected region. Calculation for the curve in Figure 6.5 takes 27 days using 4 CPU cores. Due to the long calculation time needed for theses calculations, the influence of thermionic emission from the high region on long time operation is still under calculation. However, it is supposed that its influence could be covered by the back reflected electrons from the main electron beam.

6.3 Suppression of electron beam halo and trapped electrons

One of the main limitations on the gyrotron efficiency is the maximum achievable average pitch factor α. The increase of α will lead to the increase of trapped electrons from the main beam by the magnetic mirror. According to the investigation in Chapter 5 and this chapter, it is shown that the accumulation of trapped electrons in the gyrotron can result in Low Frequencry Oscillations (LFOs) and EBHs. By understanding the generation mechanism of LFOs and EBHs in this work three methods to suppress the EBHs, LFOs and, more generally, the back reflected electrons will be presented.

6.3.1 Decrease the pitch factor at the cavity centre of electrons emitted from the neighbouring part of the emitter

As shown in the above sections, the back reflected region on the cathode surface is critical for the generation of EBH. The first method to suppress EBH is to optimize the cathode geometry which can decrease the pitch factor α value at the cavity centre of electrons emitted from neighbouring parts of the emitter [PPZ+16]. This approach eliminates the back reflected region from the cathode surface. Figure 6.12(a) shows the optimized cathode design (MIG-2) and the pitch factor at the cavity versus the axial position of the emission point of the EU 1 MW, 170 GHz ITER gyrotron [PPZ+16]

A tilt is introduced at the rear part of the cathode, as shown in Figure 6.12(a). The advantages of this tilt are as follows: (i) It decreases the local electric field around the tilt and (ii) decreases the angle between the vectors of the electric and magnetic field. According to equation 1.11 and equation 1.12, the decrease of the electric field and the angle results in the decrease of the pitch factor α at the cavity centre. Figure 6.12(b) is the "cold" pitch factor α at the cavity versus the axial position of the emitting point (blue line in Figure 6.12(a)) of the MIG-2 design for the EU 1 MW, 170 GHz ITER gyrotron. It is shown that, compared to Figure 6.1(b), the back reflected region is removed from the cathode surface in this new optimized design.

To verify this suppression method for the generation of EBH, PIC simulations have been performed using the MIG-2 design. The simulation set up is the same as the one described for the generation of EBH in the MIG-1 design, as introduced in Section 6.1.2. The distribution of electrons in the MIG-2 design at 5.0 μs is shown in Figure 6.13.

In the calculations on the MIG-1 design (Figure 6.4), EBH were generated within 1.0 μs. However, as shown in Figure 6.13, no EBH is generated in the MIG-2 design within the extended operation time (5.0 μs). In the MIG-2 design, back reflected electrons can also bombard the cathode surface and generate secondary electrons. However, due to the low α value at the cavity centre of electrons emitted from the neighbouring part, secondary electrons can not accumulate in the gyrotron MIG. Therefore, no EBH can be generated at the end.

With this optimized design of the MIG, the EU 1 MW, 170 GHz ITER gyrotron has been successfully operated with excess of 1 MW measured RF power and 34 % electronic efficiency in IHM [PPZ+16].

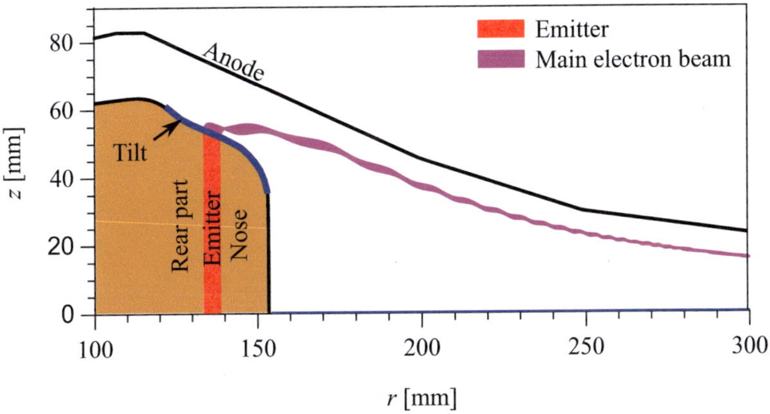

(a) Cathode-anode geometry

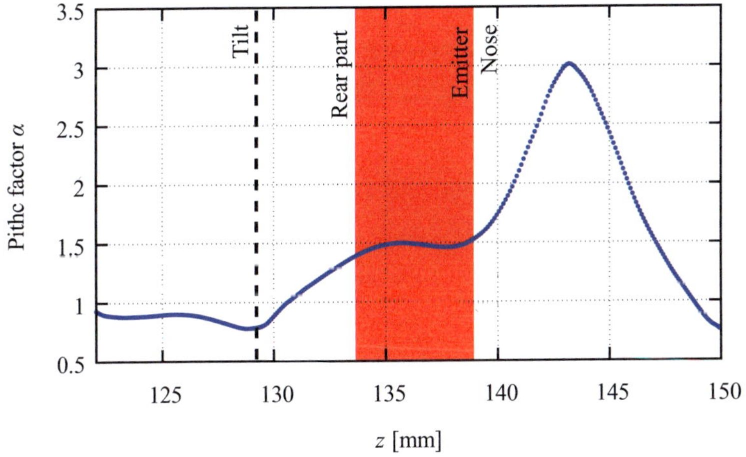

(b) The "cold" pitch factor α at the cavity centre

Figure 6.12: Sketch of (a) the cathode-anode geometry and (b) the "cold" pitch factor α at the cavity versus the axial position of the emitting point of the MIG-2 design for the EU 1 MW, 170 GHz ITER gyrotron.

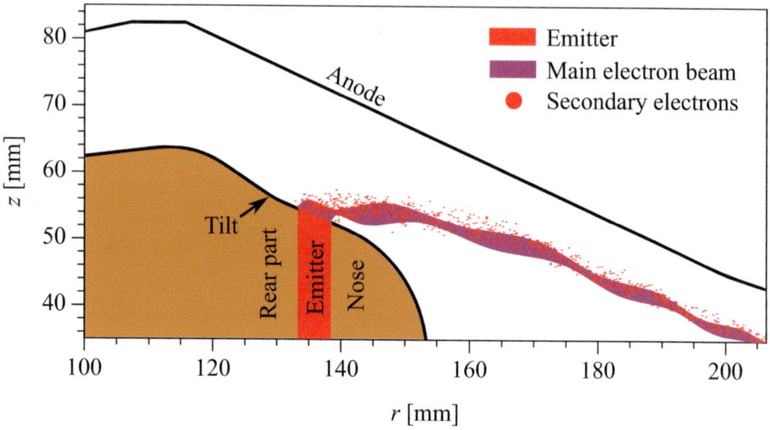

Figure 6.13: ESPIC simulation result of the distribution of electrons in the MIG-2 design for the EU 1 MW, 170 GHz ITER gyrotron at 5.0 μs.

6.3.2 Absorb the electrons emitted from the back reflected region

Another method which can suppress the generation of EBH is to absorb the secondary electrons emitted from the back reflected region. Once they are absorbed, these secondary electrons can not accumulate in the gyrotron MIG any more. Therefore, no EBH can be generated at the end. This approach has been successful applied to the refurbished first industrial prototype of the EU 2 MW, 170 GHz coaxial cavity gyrotron, as shown in Figure 6.14.

Figure 6.14(a) shows that a halo shield is introduced to the anode. The trajectories (magenta) of electrons emitted from the emitter are in a very close vicinity to the halo shield. As illustrated with the red line, the secondary electrons emitted from the rear part of the cathode will be absorbed by the halo shield.

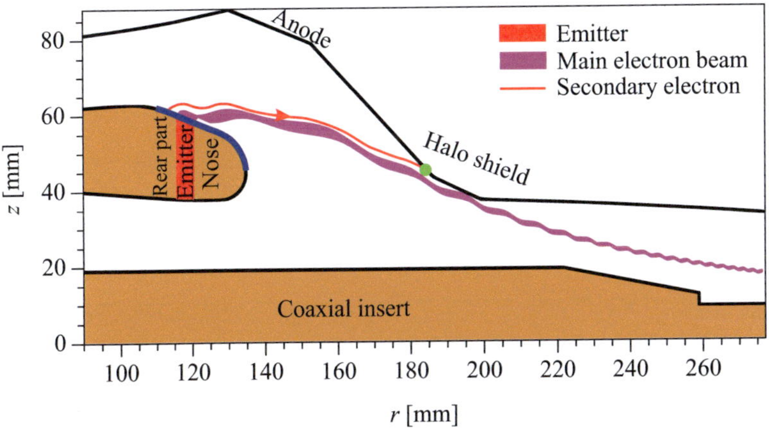

(a) Cathode-anode geometry

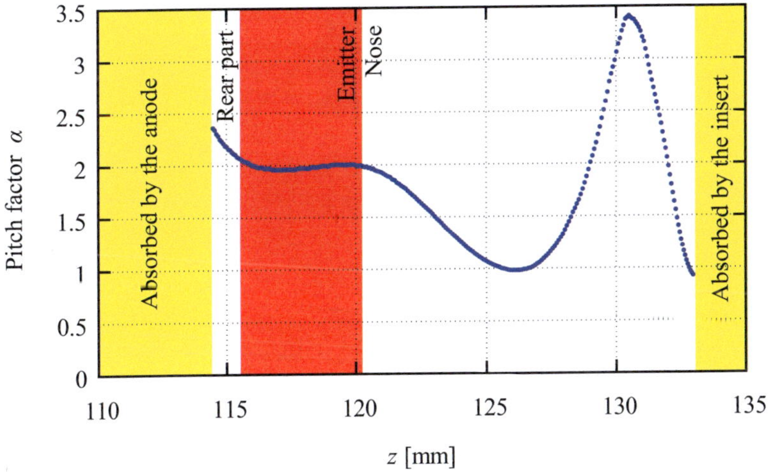

(b) The "cold" pitch factor α at the cavity centre

Figure 6.14: Sketch of (a) the cathode-anode geometry and (b) the "cold" pitch factor α at the cavity versus the axial position of the emitting point of the refurbished first industrial prototype of the EU 2 MW, 170 GHz coaxial cavity gyrotron.

In addition, a calculation using ESRAY also shows that part of the electrons emitted from the cathode nose are absorbed by the coaxial insert. Figure 6.14(b) shows the detailed information of the "cold" pitch factor α at the cavity versus the axial position of the emitting point (blue line in Figure 6.14(a)) on the cathode surface. In comparison to Figure 6.6(b) it is shown that electrons emitted from the back reflected region of the cathode surface will be absorbed by the halo shield and the coaxial insert. Therefore, the EBH shown in Figure 6.8 can not be generated in the refurbished gyrotron. To verify this concept, PIC simulations using the code ESPIC are also made for the refurbished design, as shown in Figure 6.15.

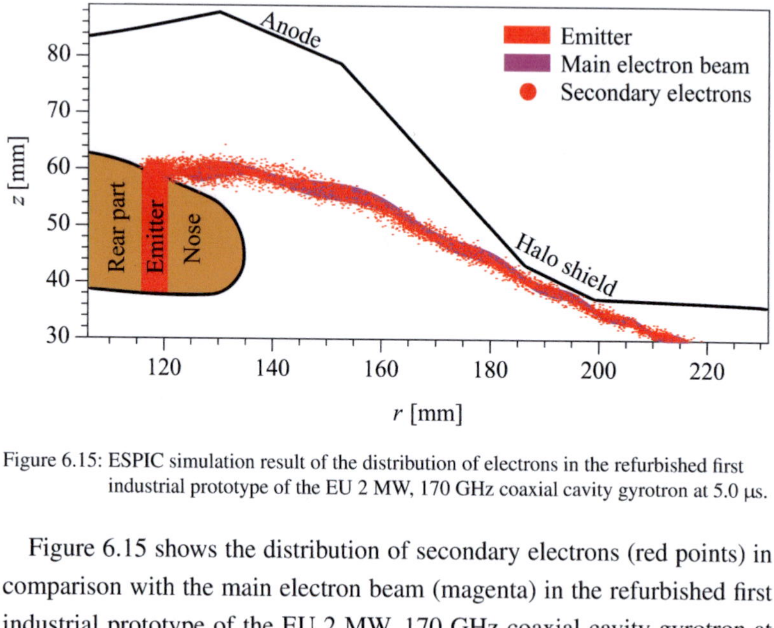

Figure 6.15: ESPIC simulation result of the distribution of electrons in the refurbished first industrial prototype of the EU 2 MW, 170 GHz coaxial cavity gyrotron at 5.0 μs.

Figure 6.15 shows the distribution of secondary electrons (red points) in comparison with the main electron beam (magenta) in the refurbished first industrial prototype of the EU 2 MW, 170 GHz coaxial cavity gyrotron at 5.0 μs. Due to the absorption of secondary electrons by the halo shield no EBH is generated. A body current of a few of mA was measured during the operation of this gyrotron which verifies the absorption of secondary electrons.

6.3.3 Extraction of back reflected electrons

Besides the above two methods a more general method to increase the electron beam quality is to extract back reflected electrons from the main beam, since they cause EBHs and LFOs of space charge in the gyrotron MIG. By extracting them, it is possible to suppress the back reflected and secondary electrons, thereby increasing the electron pitch factor, gyrotron efficiency, and output power of a gyrotron [VGGP12]. An extraction structure for the back reflected electrons is shown in Figure 6.16 using the KIT 15 kW, 28 GHz gyrotron [MIP+13] as an example.

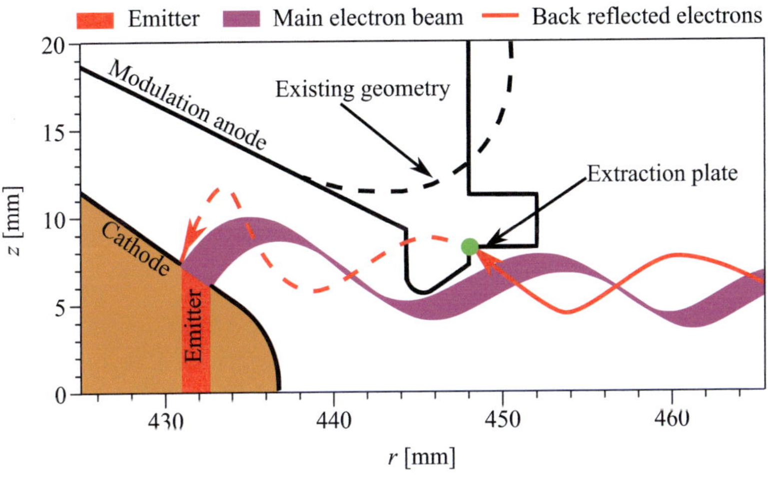

Figure 6.16: Extraction of back reflected electrons in the KIT 15 kW, 28 GHz gyrotron.

The dashed black curve is the initial structure without the extraction plate. As shown in Figure 6.16, if an electron is back reflected (red curve in Figure 6.16) from the magnetic mirror, it will not travel along the trajectory of the main electron beam. This back reflected electron will bombard the extraction plate (green point) instead of travelling back to the cathode surface. The beam parameters with and without the extraction plate are shown in Table 6.2.

Table 6.2: Comparison of beam parameters with and without the extraction plate.

Type	α	$\beta_\perp$	$\delta_{\beta_\perp}$ [%]	E_{kin}	$\delta_{E_{\mathrm{kin}}}$ [%]	r_{gc}	$\delta_{r_{\mathrm{gc}}}$ [%]
Without	1.969	0.237	2.353	19.215	0.563	3.169	18.214
With	1.998	0.237	3.605	19.220	0.4945	3.174	18.640

After adding the extraction plate, the average pitch factor α at the cavity centre of the electron beam increases slightly from 1.969 to 1.998. Both the transverse velocity spread $\delta_{\beta_\perp}$ and the guiding centre spread $\delta_{r_{\mathrm{gc}}}$ increase while the energy spread $\delta_{E_{\mathrm{kin}}}$ decreases. However, the change in the beam parameters is very small. To show the effect of the extraction plate, PIC simulation using the code ESPIC were carried out. The simulation results of the distribution of back reflected electrons with and without the extraction plate are shown in Figure 6.17.

The surface roughness model derived in Chapter 2 is used for the purpose of generating back reflected electrons. To make the extraction effect more obvious, the microstructure size of surface roughness is set to 50 μm in the simulation. Therefore, sufficient number of back reflected electrons are generated in the simulation.

In Figure 6.17, the red points are the back reflected electrons in the MIG. All of them travel in the -z direction. One can see from Figure 6.17(a) that all the back reflected electrons bombard the cathode surface, while in Figure 6.17(b), most of them bombard the extraction plate. According to the analysis of the simulation result, only 10 % of back reflected electrons can bombard the cathode surface.

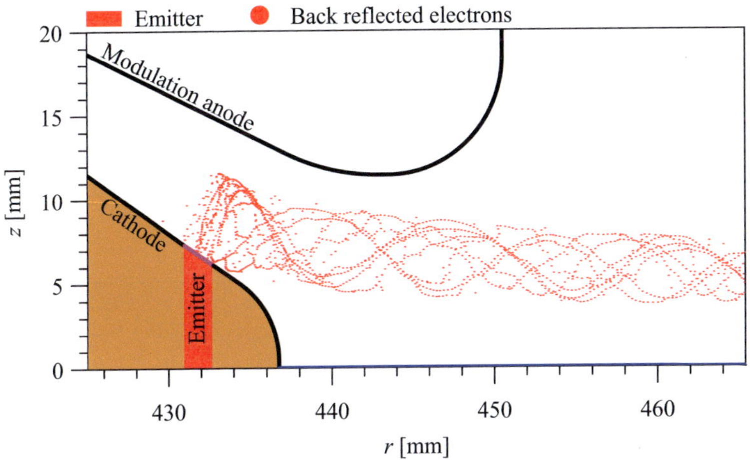

(a) Without the extraction plate

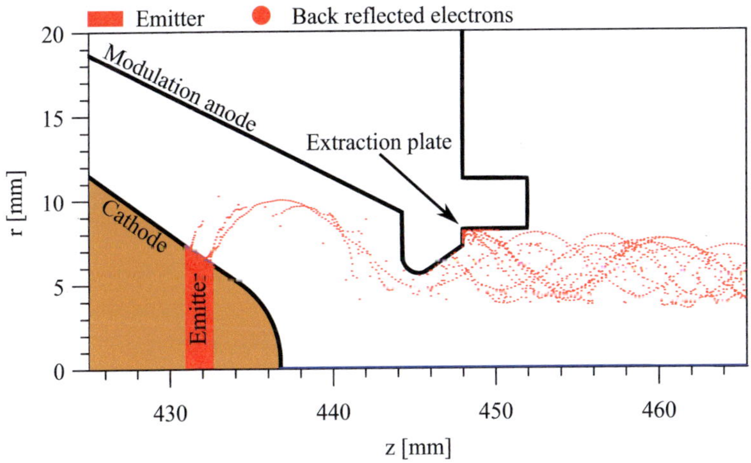

(b) With the extraction plate

Figure 6.17: Distribution of back reflected electrons (a) without and (b) with the extraction plate.

Besides stopping back reflected electrons from bombarding the cathode surface the extraction plate can also decrease the number of secondary electrons in the gyrotron. Figure 6.18 shows the space charge of secondary electrons between the cathode and the cavity centre.

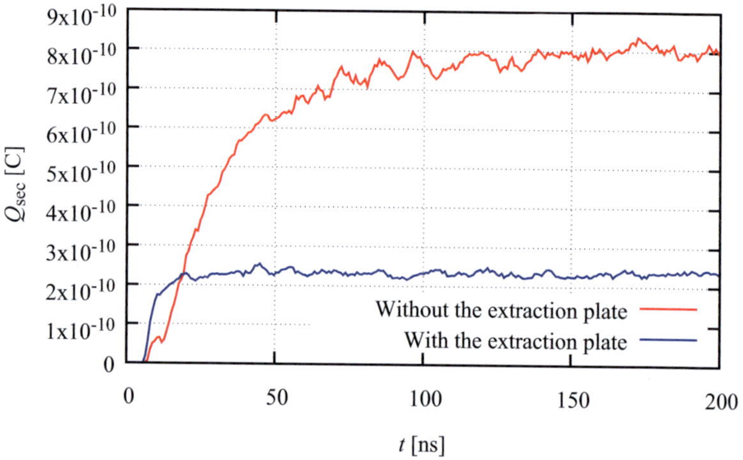

Figure 6.18: Space charge of secondary electrons between the cathode and the cavity centre without and with the extraction plate.

As shown in Figure 6.18, with the extraction plate, the space charge of secondary electrons between the cathode and the cavity is decreased to 2.2×10^{-10}C from 8×10^{-10}C which is around four times smaller. This is due to two facts. Firstly, the extraction plate generates less secondary electrons. Secondly, the extraction plate absorbs most of the secondary electrons. The kinetic energy of the back reflected electrons when they bombard the cathode surface is between 0.5 keV and 1 keV while the kinetic energy of the back reflected electrons when they bombard the extraction plate is approximately 15.6 keV. According to the work described in Chapter 4, the Secondary Emission Yield (SEY) of tungsten when bombarded by 1 keV electron beam is approximately 2.0 while the SEY for copper is only 0.95 when bombarded by 15.6 keV electron beam. Therefore, less secondary electrons are generated with the extraction plate. In addition, most of the secondary electrons emitted from the cathode surface will be trapped between the cathode and the cavity centre (Section 5.1) but most secondary electrons emitted from the extraction plate will be either absorbed by the absorption structure or travel out through the cavity. The extraction plate works like a collector for the back reflected electrons and secondary electrons. At the end, the number of secondary electrons in the gyrotron MIG can be significantly decreased. Therefore, such an extraction plate can absorb most of the back reflected electrons and decrease the number of secondary electrons in the gyrotron without significantly changing the electron beam parameters.

It shall be pointed out that the extraction plate is more suitable in the case where there is a large Larmor radius for the electrons above the cathode surface, since in this case there is more space and, therefore, more flexibility for the design of a extraction plate. High power gyrotrons usually have bigger emitter rings due to the high beam current requirement. The Larmor radius is relatively small compared to the emitter ring radius in these gyrotrons. Therefore, the flexibility of using the extraction plate in these gyrotrons is quite limited and the two methods discussed in Section 6.3.1 and Section 6.3.2 are more suitable for this purpose.

6.4 Summary

The generation mechanism of harmful EBHs is presented and discussed. Numerical investigation is carried out using the enhanced code ESPIC to demonstrate the mechanism. The generation of EBHs in two gyrotron MIGs are investigated. The correlation between damaged spots on the surface of gyrotron components and EBH is also verified by the calculation using ESPIC. The possibility that an EBH is generated by thermionic emitted electrons from the back reflected region is investigated. Due to the significantly less space charge of the trapped electrons, it is supposed that its influence will be covered by the back reflected electrons from the main electron beam.

By understanding the generation mechanism of LFOs and EBHs in this work three methods to suppress an EBH and, more generally, the trapped electrons are presented. It is numerically proven that eliminating the back reflected region from the cathode surface or absorbing electrons emitted from this region can totally suppress the generation of an EBH. New gyrotron MIGs based on these two optimization methods have been successfully operated at IHM. A novel MIG geometry with a extraction plate is proposed for the general suppression of back reflected electrons and secondary electrons in gyrotron MIGs. The simulations with the codes ESRAY and ESPIC show that the extraction plate can decrease the number of back reflected electrons which bombard the cathode surface by 90 % without significantly changing the beam parameters. With the extraction plate, the number of secondary electrons between the cathode and the cavity centre is reduced by a factor of four.

7 Conclusion and outlook

7.1 Summary

The influences of two adverse factors: emitter surface roughness and emission inhomogeneity on gyrotron operation were investigated with a focus on Low Frequency Oscillations (LFOs) and Electron Beam Halo (EBH). To achieve this, an emitter surface roughness model (Chapter 2), a new evaluation method for the emission inhomogeneity (Chapter 3) and a modified secondary electron emission model (Chapter 4) were developed. The surface roughness model and secondary electron emission model were implemented in the KIT in-house beam optics codes ESRAY and ESPIC. LFOs and EBH were investigated using the enhanced code and compared with the experimental results. By understanding these mechanisms optimization methods to suppress these two harmful phenomena are presented.

A new surface roughness model is derived for the KIT in-house codes ESRAY and ESPIC in order to investigate the influence of surface roughness on electron beam quality and gyrotron efficiency. The surface roughness model uses an initial velocity distribution to simulate the influence of surface roughness. Thus, one can calculate the influence of surface roughness without modeling real surface microstructures and increase the mesh number in the code. This feature decreases the calculation time significantly. The influences of surface roughness on the electron beam quality and gyrotron efficiency of two different versions of the MIG for the EU 1 MW, 170 GHz ITER gyrotron were calculated using the codes ESRAY and EURIDICE. The increase of surface roughness from 0 µm to

25 µm results in a decrease of the total efficiency (theoretical value with single-stage depressed collector) η_{tot} by 15 % (from 70.5 % to 55.5 %) and 9 % (from 69 % to 60 %) respectively for these two versions of the gyrotron. Multimode calculations using the code EURIDICE show that surface roughness can suppress weak competing modes and facilitates mode selection. The electron beam calculation using the code ESRAY shows that back reflected electrons can be generated if the microstructure size is larger than 17 µm for the short pulse operation.

In order to investigate the influence of emission inhomogeneity, a novel definition of emission inhomogeneity δ_J and its evaluation method is derived. Comparison with the former definitions shows that the new definition is intuitive and stable. The emission inhomogeneities of three different gyrotrons are measured using the new definition. The measurement results indicate that the emission inhomogeneity increases with the increase of operating time and number of tube openings. The influence of emission inhomogeneity on the electron beam quality and gyrotron efficiency is calculated using the 3D code ARIADNE and EURIDICE, respectively. The calculation results show that $\delta_J = 0.3$ can result in a decrease in gyrotron efficiency by approximately 2 % (e.g., from 41.38 % to 39.46 % for the EU 1 MW, 170 GHz ITER gyrotron). The influence of work function, temperature and electric field distribution on the emission inhomogeneity is numerically calculated using a Monte Carlo method. To keep the influence of emission inhomogeneity on gyrotron efficiency lower than 2%, the boundary values for the distribution of work function, temperature and electric field are $\sigma_T < 20\,°\mathrm{C}$, $\sigma_W < 0.03\,\mathrm{eV}$ and $\delta_E < 0.5$. The electron beam calculation using the 3D code ARIADNE shows that back reflected electrons can be generated due to emission inhomogeneity when operating at high values of the pitch factor α.

To investigate the influence of electrons back reflected by the magnetic mirror, a modified Furman secondary electron emission model is implied in the codes ESRAY and ESPIC. The modified model has a better accuracy for the angular distribution of high energy secondary electrons (elastic and rediffused electrons). The secondary electron emission parameters of tungsten and molybdenum, which are used in the gyrotron cathode, are derived using the fitting method. With this new secondary electron emission model, it is possible to calculate the influence of back reflected electrons in the gyrotron MIG using the code ESPIC.

One of the phenomena caused by the back reflected electrons is space charge LFOs. Particle In Cell (PIC) calculations using the code ESPIC show that an 140 MHz space charge LFO in the EU 2 MW, 170 GHz coaxial cavity gyrotron can be initiated. The space charge LFO results from the charge bunching of the trapped electrons between the cathode and magnetic mirror. The beam-wave instability between the electron beam and TEM mode is presented. Two TEM mode LFOs (64.9 MHz and 114.8 MHz) have been observed during the experiment. The measured two TEM modes have the same characteristic as described in the theoretical analysis. The 64.9 MHz TEM mode has a higher amplitude due to higher beam-wave coupling and low radiation loss. The 114.8 MHz TEM mode has a wider exciting range due to the correlation with the 140 MHz space charge LFO. One possible solution to suppress the TEM mode is to move the emitter position to the left of the point where $|\vec{E}_{\mathrm{TEM}}| = 0$. In this case the TEM mode will lose energy to the electron beam in one period, therefore, it can not be excited.

Harmful EBHs can also be initiated if there is a back reflected region on the cathode surface. EBHs initiated by back reflected electrons from the main electron beam in different gyrotron MIGs are investigated using the code ESPIC. The calculation verifies the correlation between damaged spots on the surface of gyrotron components and the EBH. By understanding the generation mechanism of LFOs and EBHs in this work three methods to suppress EBHs and, more generally, the trapped electrons are presented. It is numerically proved that elimination of the back reflected region from the cathode surface or absorption of electrons emitted from this region can totally suppress the generation of EBH. Two new optimized gyrotrons

MIGs have been successfully operated. A novel MIG geometry with a halo shield is proposed for the general suppression of back reflected electrons and secondary electrons. Simulations with the codes ESRAY and ESPIC show that the halo shield can decrease the number of the back reflected electrons which bombard the cathode surface by $90\,\%$ without significantly changing the beam parameters. With the halo shield, the number of secondary electrons between the cathode and the cavity center is also around 4 times smaller.

In the current state, the enhanced beam optics codes are new and very valuable tools to understand and analyze the deteriorating phenomena. By understanding the mechanisms of harmful LFOs and EBHs using these enhanced codes, optimization methods are presented in this work. Hence, the enhanced codes and the mechanisms studied in this work are important to assist in the advanced design of more robust magnetron injection guns for gyrotrons in the future.

7.2 Outlook

In the following part, the most obvious next steps for a real time emission inhomogeneity measurement system, a beam-wave instability simulation and the design of gyrotron MIGs with extraction plate are listed.

Real time emission inhomogeneity measurement system

In this work, it has been verified that emission inhomogeneity can decrease electron beam quality and gyrotron efficiency by increasing the velocity, energy, and guiding center spread of the electron beam. The emission inhomogeneity is an important factor to evaluate the beam quality. Real time measurement of the emission inhomogeneity during the experiment can supply additional useful information for the researchers. A code has been developed for the evaluation of emission inhomogeneity from the Current Voltage Characteristics (CVC). However, due to the limitations of the current high power supply, it is impossible to perform real time measurements. For the next generation of a gyrotron teststand, which is currently under construction at the Institute for Pulsed Power and Microwave Technology (IHM), a real time emission inhomogeneity measurement system can be implied without the requirement for any additional devices. First, this system can be used to make a qualification of the MIG once a new gyrotron is constructed. Then, with this system the change of emission inhomogeneity during the emitter conditioning can be observed during the experiment. The emitter aging process can also be studied during the long period (several years) operation. Furthermore, this system can also be used to check the beam quality after an electric breakdown has occurred between the cathode and anode.

Simulation of low frequency beam-wave interaction

The beam-wave instability in the EU 2 MW, 170 GHz coaxial cavity short-pulse prototype gyrotron has been qualitatively analyzed using a simplified resonator model. Due to the wide exciting range of the 114.8 MHz TEM mode, which is measured during the experiment, it was pointed out that the 140 MHz space charge LFO can be an excitation source for this TEM mode. Verification of this phenomenon needs beam-wave interaction simulations. Initial calculations show that by adding a TEM mode to the gyrotron MIG, the space charge LFO is locked to the frequency of the TEM mode. The beam parameters also have a periodical variation with the TEM mode. However, at present, full beam-wave interaction features have not been implemented into ESPIC. Due to the current limitations, it can only calculate the influence of the TEM mode on the electrons but can not take into account the inverse influence. Therefore, the amplitude of the TEM mode is constant during the simulation. The next step is to take into account the full beam-wave interaction effect.

Design of gyrotron MIG with extraction plate for back reflected electrons

As shown in the last two chapters, back reflected electrons resulting from surface roughness and emission inhomogeneity can initiate LFOs and EBHs. From the design point of view for a MIG, the extraction plate is a promising method. The discussion in Section 6.3 already shows that an extraction plate can absorb the back reflected electrons without significantly changing the beam parameters. To have a wide operation range, one shall optimize the beam shape during the design of a gyrotron MIG with that extraction plate. Otherwise, the main electron beam may bombard the extraction plate directly. In future designs of low power gyrotron MIGs, the possibility of using the extraction plate shall be considered and properly designed.

Bibliography

[AKT+02] J. P. Anderson, S. E. Korbly, R. J. Temkin, M. A. Shapiro, K. L. Felch, and S. Cauffman. Design and emission uniformity studies of a 1.5-rf wave gyrotron electron gun. *Plasma Science, IEEE Transactions on*, 30(6):2117–2123, Dec 2002.

[APIV12] K. A. Avramides, I. Gr. Pagonakis, C. T. Iatrou, and J. L. Vomvoridis. Euridice: A code-package for gyrotron interaction simulations and cavity design. In *EPJ Web of Conferences*, volume 32, page 04016. EDP Sciences, 2012.

[ATS05] J. P. Anderson, R. J. Temkin, and M. A. Shapiro. Experimental studies of local and global emission uniformity for a magnetron injection gun. *Electron Devices, IEEE Transactions on*, 52(5):825–828, May 2005.

[Ber11] Matthias Hermann Beringer. *Design studies towards a 4 MW 170 GHz coaxial-cavity gyrotron*. PhD thesis, Karlsruhe Institute of Technology, Karlsruhe, 2011. KIT Scientific Publishing, ISBN 978-3-86644-663-2.

[BH97] E. Borie and U. Horcher. Effect of surface roughness on the velocity spread in electron guns for gyrotrons. *International Journal of Infrared and Millimeter Waves*, 18(3):577–594, 1997.

[BL97] D. V. Borzenkov and O. I. Louksha. Numerical simulation of space-charge dynamics in a gyrotron trap. *Technical Physics*, 42(9):1071–1074, 1997.

[BLC+08] L. R. Barnett, N. C. Luhmann, C. C. Chiu, Y. C. Yan, and K. R. Chu. An improved magnetron injection gun using advanced high current density cathodes for a w-band te01 gyro-twt. In *Vacuum Electronics Conference, 2008. IVEC 2008. IEEE International*, pages 400–401, April 2008.

[Bun67] O. Buneman. Time-reversible difference procedures. *Journal of Computational Physics*, 1(4):517 – 535, 1967.

[CCM+09] A. J. Cerfon, E. Choi, C. D. Marchewka, I. Mastovsky, M. A. Shapiro, J. R. Sirigiri, and R. J. Temkin. Observation and study of low-frequency oscillations in a 1.5-MW 110-GHz gyrotron. *Plasma Science, IEEE Transactions on*, 37(7):1219–1224, July 2009.

[Che74] F. F. Chen. *Introduction to plasma physics*. Plenum Press, New York, 1974.

[Che16] F. F. Chen. *Introduction to Plasma Physics and Controlled Fusion*. Springer International Publishing, 2016.

[Chi11] C. D. Child. Discharge from hot CaO. *Phys. Rev. (Series I)*, 32:492–511, May 1911.

[Chu04] K. R. Chu. The electron cyclotron maser. *Rev. Mod. Phys.*, 76:489–540, May 2004.

[Chu15] T. Chu. Microwave circuits. University Lecture, 2015.

[Cro65] C. R. Crowell. The richardson constant for thermionic emission in schottky barrier diodes. *Solid-State Electronics*, 8(4):395 – 399, 1965.

[Cro81] J. L. Cronin. Modern dispenser cathodes. *IEE Proceedings I (Solid-State and Electron Devices)*, 128(1):19–32, 1981.

[DCJ+07] D. Drouin, A. R. Couture, D. Joly, X. Tastet, V. Aimez, and R. Gauvin. CASINO v2. 42: A fast and easy-to-use modeling tool for scanning electron microscopy and microanalysis users. *Scanning*, 29(3):92–101, 2007.

[DK81] A. T. Drobot and K. Kim. Space charge effects on the equilibrium of guided electron flow with gyromotion. *International Journal of Electronics*, 51(4):351–367, 1981.

[DP97] O. Dumbrajs and A. B. Pavelyev. Insert misalignment in coaxial cavities and its influence on gyrotron operation. *International Journal of Electronics*, 82(3):261–268, 1997.

[DPDC+11] H. Demers, N. Poirier-Demers, A. R. Couture, D. Joly, M. Guilmain, N. de Jonge, and D. Drouin. Three-dimensional electron microscopy simulation with the CASINO monte carlo software. *Microscopy and Microanalysis*, 17:612–613, 7 2011.

[Edg93] C. J. Edgcombe, editor. *Gyrotron oscillators: their principles and practice*. Taylor & Francis, 1993.

[Edg95] C. J. Edgcombe. Sources of velocity spread in electron beams from magnetron injection guns. *International Journal of Infrared and Millimeter Waves*, 16(1):83–97, 1995.

[Ein00] A. Einstein. Elementary derivation of the equivalence of mass and energy. *Bulletin of the American Mathematical Society*, 37(1):39–44, 2000.

[EK81] G. Eng and H. K. A. Kan. Scanning auger and work-function measurements applied to dispenser cathodes. *Applications of Surface Science*, 8(1):81 – 94, 1981.

[Fal83] L. R. Falce. Dispenser cathodes: the current state of the technology. In *Electron Devices Meeting, 1983 International*, volume 29, pages 448–451. IEEE, 1983.

[FEV03] Richard G Forbes, C.J Edgcombe, and U ValdrÃ¨. Some comments on models for field enhancement. *Ultramicroscopy*, 95:57 – 65, 2003. {IFES} 2001.

[FHJ90] K. Felch, H. Huey, and H. Jory. Gyrotrons for ECH applications. *Journal of Fusion Energy*, 9(1):59–75, 1990.

[Fla12] J. H. Flamm. *Diffraction and Scattering in Launchers of Quasi-Optical Mode Converters for Gyrotrons*. PhD thesis, Karlsruhe Institute of Technology, Karlsruhe, 2012.

[FP02] M. Furman and M. Pivi. Probabilistic model for the simulation of secondary electron emission. *Physical Review Special Topics - Accelerators and Beams*, 5(12):124404, December 2002.

[GEI+10] G. Gantenbein, V. Erckmann, S. Illy, S. Kern, W. Kasparek, C. Lechte, W. Leonhardt, C. Liévin, A. Samartsev, A. Schlaich, M. Schmid, and M. Thumm. 140 GHz, 1 MW 1 CW gyrotron development for fusion applications—progress and recent results. *Journal of Infrared, Millimeter, and Terahertz Waves*, 32(3):320–328, 2010.

[GGK+99] M. Y. Glyavin, A. L. Goldenberg, A. N. Kuftin, V. K. Lygin, A. S. Postnikova, and V. E. Zapevalov. Experimental studies of gyrotron electron beam systems. *IEEE Transactions on Plasma Science*, 27(2):474–483, Apr 1999.

[GGM+10] I. G. Gachev, M. Y. Glyavin, V. N. Manuilov, M. V. Morozkin, and N. A. Zavolsky. The influence of initial electron velocities distribution on the energy spectra of the spent electron beam in gyrotron. *Journal of Infrared, Millimeter, and Terahertz Waves*, 31(10):1109–1114, 2010.

[GKM14] M. Y. Glyavin, A. D. Kuntsevich, and V. N. Manuilov. Suppression of the Oscillatory Modes of a Space Charge in the Magnetron Injection Guns of Technological Gyrotrons. *Journal of Infrared, Millimeter, and Terahertz Waves*, 36(1):7–12, 2014.

[GKV+97] M. Y. Glyavin, A. N. Kuftin, N. P. Venediktov, A. S. Postnikova, M. V. Yulpatova, and V. E. Zapevalov. Experimental investigation of emission inhomogeneity of gyrotron cathodes basing on their current-voltage characteristics. *International journal of infrared and millimeter waves*, 18(11):2137–2146, 1997.

[GLP+99] I. Gorelov, J. M. Lohr, D. Ponce, R. W. Callis, H. Izeki, R. A. Legg, and S. E. Tsimring. Gyrotron performance on the 110 GHz installation at the DIII-D tokamak. In *24th Int. Conf. Infrared Millimiter Waves*, Monterey, CA, 1999.

[Har02] A. A Harms. *Principles of fusion energy*. Allied Publishers, 2002.

[Has70] W. K. Hastings. Monte Carlo sampling methods using Markov chains and their applications. *Biometrika*, 57(1):97–109, 1970.

[HDG+09] R. Hemsworth, H. Decamps, J. Graceffa, B. Schunke, M. Tanaka, M. Dremel, A. Tanga, H. P. L. De Esch, F. Geli, J. Milnes, T. Inoue, D. Marcuzzi5, P. Sonato, and P. Zaccaria.

Status of the ITER heating neutral beam system. *Nuclear Fusion*, 49(4):045006, 2009.

[IBC+08] R. L. Ives, P. Borchard, G. Collins, M. Curtis, L. R. Falce, K. Gunther, D. Marsden, G. Miram, and R. Wilcox. Improved magnetron injection guns for gyrotrons. *IEEE Transactions on Plasma Science*, 36(3):620–630, June 2008.

[IDPH+01] T. Inoue, E. Di Pietro, M. Hanada, R. Hemsworth, A. Krylov, V. Kulygin, P. Massmann, P. L. Mondino, Y. Okumura, A. Panasenkov, E. Spethc, and K. Watanabed. Design of neutral beam system for ITER-FEAT. *Fusion engineering and design*, 56:517–521, 2001.

[ILM+98] V. N. Ilyin, O. I. Louksha, V. E. Mjasnikov, L. G. Popov, and G. G. Sominski. Effect of emission inhomogeneities on low-frequency oscillations in gyrotron-type electron beams. In *High-Power Particle Beams, 1998. BEAMS'98. Proceedings of the 12th International Conference on*, volume 2, pages 800–803, Haifa, Jun 1998. IEEE.

[ISN+98] T. Idehara, K. Shibutani, H. Nojima, M. Pereyaslavets, K. Yoshida, I. Ogawa, and T. Tatsukawa. Study of electron beam misalignment in a submillimeter wave gyrotron. *International Journal of Infrared and Millimeter Waves*, 19(10):1303–1316, 1998.

[IZJ15] S. Illy, J. Zhang, and J. Jelonnek. Gyrotron electron gun and collector simulation with the ESRAY beam optics code. In *Vacuum Electronics Conference, IEEE International*, pages 1–2, April 2015.

[JAA+13] J. Jelonnek, S. Alberti, K. A. Avramidis, V. Erckmann, G. Gantenbein, K. Hesch, J. P. Hogge, S. Illy, J. Jin, S. Kern,

I. Gr. Pagonakis, B. Piosczyk, T. Rzesnicki, A. Samartsev, and M. Thumm. Development of advanced gyrotrons in EUROPE. *Fusion Science and Technology*, 64(3):505–512, 2013.

[JAF+13] J. Jelonnek, K. Avramidis, J. Franck, G. Gantenbein, K. Hesch, S. Illy, J. Jin, A. Malygin, I. Gr. Pagonakis, T. Rzesnicki, A. Samartsev, T. Scherer, A. Schlaich, M. Schmid, D. Strauss, M. Thumm, and J. Zhang. KIT gyrotron development for future fusion applications. In *Infrared, Millimeter, and Terahertz Waves (IRMMW-THz), 2013 38th International Conference on*, pages 1–2, Sept 2013.

[JAF+14] J. Jelonnek, K. Avramidis, J. Franck, G. Dammertz, G. Gantenbein, K. Hesch, S. Illy, J. Jin, A. Malygin, I. Gr. Pagonakis, B. Piosczyk, T. Rzesnicki, A. Samartsev, A. Schlaich, M. Thumm, and J. Zhang. KIT contribution to the gyrotron development for nuclear fusion experiments in europe. In *Microwave Conference (GeMIC), 2014 German*, pages 1–4, March 2014.

[JC71] D. Jacobson and A. Campbell. Molybdenum work function determined by electron emission microscopy. *Metallurgical and Materials Transactions B*, 2(11):3063–3066, 1971.

[Jel00] J. Jelonnek. *Untersuchung des Lastverhaltens von Gyrotrons*. PhD thesis, Technical University Hamburg-Harburg, Düsseldorf, 2000.

[JGH+13] J. Jelonnek, G. Gantenbein, K. Hesch, J. Jin, I. Gr. Pagonakis, B. Piosczyk, T. Rzesnicki, M. Thumm, S. Alberti, J.-P. Hogge, M. Q. Tran, V. Erckmann, H. Laqua, G. Michel, P. Benin, F. Legrand, Y. Rozier, K. Avramidis, J. L. Vomvoridis, Z. C. Ioannidis, G. P. Latsas, I. G. Tigelis, F. Albajar, T. Bonicelli, and F. Cismondi. From series production of gyrotrons for W7-X towards EU-1 MW gyrotrons for ITER. In *Pulsed Power Conference (PPC), 2013 19th IEEE*, pages 1–6, June 2013.

[Jin07] J. Jin. Quasi-optical mode converter for a coaxial cavity gyrotron. Technical report, Forschungszentrum Karlsruhe, Karlsruhe, 2007. Wissenschaftliche Berichte des Forschungszentrum Karlsruhe FZKA 7264.

[Joy95] D. C. Joy. A database on electron-solid interactions. *Scanning*, 17(5):270–275, 1995.

[JTP+09] J. Jin, M. Thumm, B. Piosczyk, S. Kern, J. Flamm, and T. Rzesnicki. Novel numerical method for the analysis and synthesis of the fields in highly oversized waveguide mode converters. *IEEE Transactions on Microwave Theory and Techniques*, 57(7):1661–1668, July 2009.

[Kam75] T. Kammash. *Fusion reactor physics: principles and technology*. Ann Arbor Science, 1975.

[Kan57] H. Kanter. Zur rückstreuung von elektronen im energiebereich von 10 bis 100 kev. *Annalen der Physik*, 455(1-6):144–166, 1957.

[KB08] E. Kieft and E. Bosch. Refinement of Monte Carlo simulations of electron-specimen interaction in low-voltage SEM. *Journal of Physics D: Applied Physics*, 41(21):215310, 2008.

[KDST85] K. E. Kreischer, B. G. Danly, J. B. Schutkeker, and R. J. Temkin. The design of megawatt gyrotrons. *Plasma Science, IEEE Transactions on*, 13(6):364–373, Dec 1985.

[Ker96] S. Kern. *Numerische Simulation der Gyrotron-Wechselwirkung in koaxialen Resonatoren*. PhD thesis,

Universität Karlsruhe, Juli 1996. Wissenschaftliche Berichte des Forschungszentrums Karlsruhe FZKA 5837.

[Kho69] S. P. Khodnevich. Determination of emission homogeneity on the cathode by current-voltage characteristic. *Electron. Technol., Microwave Electron.*, 20(4):118 – 130, 1969.

[KLP+04] D. V. Kas'yanenko, O. I. Louksha, B. Piosczyk, G. G. Sominsky, and M. Thumm. Low-frequency parasitic space-charge oscillations in the helical electron beam of a gyrotron. *Radiophysics and Quantum Electronics*, 47(5):414–420, 2004.

[Kun80] W. B. Kunkel. *Fusion*, volume 1, Part B (Edited by E. Teller). Academic Press, 1980.

[Kup79] A. Kupiszewski. The gyrotron: A high-frequency microwave amplifier. *The Deep Space Network Progress Report 42*, 52:8, 1979.

[LAL96] C. Liu, T. M. Antonsen, and B. Levush. Simulation of the velocity spread in magnetron injection guns. *Plasma Science, IEEE Transactions on*, 24(3):982–991, Jun 1996.

[Lau87] Y. Y. Lau. Effects of cathode surface roughness on the quality of electron beams. *Journal of Applied Physics*, 61(1):36–44, Jan 1987.

[Law57] J. D. Lawson. Some Criteria for a Power Producing Thermonuclear Reactor. *Proceedings of the Physical Society B*, 70:6–10, January 1957.

[Lon03] R. T. Longo. Physics of thermionic dispenser cathode aging. *Journal of Applied Physics*, 94(10):6966–6975, 2003.

[Lou09] O. I. Louksha. Simulation of low-frequency collective processes in gyrotron electron beams. *Radiophysics and Quantum Electronics*, 52(5-6):386–397, 2009.

[LPS+06a] O. I. Louksha, B. Piosczyk, G. Sominski, M. Thumm, and D. Samsonov. Electron emission inhomogeneity and low-frequency parasitic oscillations in a gyrotron. *AIP Conference Proceedings*, 807:219–220, 2006.

[LPS+06b] O. I. Louksha, B. Piosczyk, G. G. Sominski, M. K. Thumm, and D. B. Samsonov. On potentials of gyrotron efficiency enhancement: measurements and simulations on a 4-mm gyrotron. *Plasma Science, IEEE Transactions on*, 34(3):502–511, June 2006.

[LS12] M. C. Lin and D. N. Smithe. 3D CFDTD PIC simulation study on low-frequency oscillations in a gyrotron. In *2012 IEEE 13th International Vacuum Electronics Conference, IVEC 2012*, pages 541–542, 2012.

[LSK99] O. I. Louksha, G. G. Sominski, and D. V. Kas'yanenko. Experimental study and numerical simulation of electron beam in gyrotron-type electron-optical systems. In *Electronics and Radiophysics of Ultra-High Frequencies, 1999. International University Conference Proceedings*, pages 130–133, Aug 1999.

[LST11] A. Litvak, K. Sakamoto, and M. Thumm. Innovation on high-power long-pulse gyrotrons. *Plasma Physics and Controlled Fusion*, 53(12):124002, 2011.

[Lyg95] V. K. Lygin. Numerical simulation of intense helical electron beams with the calculation of the velocity distribution functions. *International journal of infrared and millimeter waves*, 16(2):363–376, 1995.

[LYG04] Z. Li, J. Yu, and B. Gao. *Foundations for microwave engineering*. Tsinghua University press, 2004.

[Man09] V. N. Manuilov. Space charge oscillations in gyrotron magnetron-injection guns with high pitch-factor. *2009 IEEE International Vacuum Electronics Conference*, 1:90–91, April 2009.

[MG56] E. L. Murphy and R. H. Good. Thermionic emission, field emission, and the transition region. *Phys. Rev.*, 102:1464–1473, Jun 1956.

[MIP+13] A. Malygin, S. Illy, I. Gr. Pagonakis, B. Piosczyk, S. Kern, J. Weggen, M. Thumm, J. Jelonnek, K. A. Avramides, R. L. Ives, D. Marsden, and G. Collins. Design and 3-D simulations of a 10-kW/28-GHz gyrotron with a segmented emitter based on controlled porosity-reservoir cathodes. *Plasma Science, IEEE Transactions on*, 41(10):2717–2723, 2013.

[MIR+04] G. Miram, L. Ives, M. Read, R. Wilcox, M. Cattelino, and B. Stockwell. Emission spread in thermionic cathodes. In *Vacuum Electronics Conference, 2004. IVEC 2004. Fifth IEEE International*, pages 303–304, April 2004.

[MMS71] K. Murata, T. Matsukawa, and R. Shimizu. Monte Carlo calculations on electron scattering in a solid target. *Japanese Journal of Applied Physics*, 10(6):678, 1971.

[MZI06] V. N. Manuilov, V. Yu. Zaslavsky, and T. Idehara. Two-dimensional numerical simulation of low frequency oscillations of space charge and potential in the gyrotron adiabatic trap. *International Journal of Infrared and Millimeter Waves*, 27(12):1573–1593, 2006.

[NGM+56] W. B. Nottingham, R. H. Good, E. W. Müller, R. Kollath, G. L. Weissler, W. P. Allis, L. B. Loeb, A. Engel, and P. F. Little. Electron-emission gas discharges I / Elektronen-emission gasentladungen I. *Handbuch der Physik*, 21, 1956.

[NVB+01] G. S. Nusinovich, A. N. Vlasov, M. Botton, T. M. Antonsen, S. Cauffman, and K. Felch. Effect of the azimuthal inhomogeneity of electron emission on gyrotron operation. *Physics of Plasmas*, 8(7):3473–3479, 2001.

[PAA+15] I. Gr. Pagonakis, F. Albajar, S. Alberti, K. A. Avramidis, T. Bonicelli, F. Braunmueller, A. Bruschi, I. Chelis, F. Cismondi, G. Gantenbein, V. Hermann, K. Hesch, J.-P. Hogge, J. Jelonnek, J. Jin, S. Illy, Z. Ioannidis, T. Kobarg, G. P. Latsas, F. Legrand, M. Lontano, B. Piosczyk, Y. Rozier, T. Rzesnicki, A. Samartsev, C. Schlatter, M. Thumm, I. G. Tigelis, M.-Q. Tran, T.-M. Tran, J. Weggen, and J. L. Vomvoridis. Status of the development of the EU 170 GHz/1 MW CW gyrotron. *Fusion Engineering and Design*, 2015.

[Pag15] I. Gr. Pagonakis. personal communication, 2015.

[PBD+99] B. Piosczyk, O. Braz, G. Dammertz, C. T. Iatrou, S. Illy, M. Kuntze, G. Michel, and M. Thumm. 165 GHz, 1.5 MW-coaxial cavity gyrotron with depressed collector. *Plasma Science, IEEE Transactions on*, 27(2):484–489, Apr 1999.

[PCE+11] R. A. Pitts, S. Carpentier, F. Escourbiac, T. Hirai, V. Komarov, A. S. Kukushkin, S. Lisgo, A. Loarte, M. Merola, R. Mitteau, A. R. Raffray, M. Shimada, and P. C. Stangeby. Physics basis and design of the ITER plasma-facing components. *Journal of Nuclear Materials*, 415(1):957–964, aug 2011.

[PDD+04] B. Piosczyk, G. Dammertz, O. Dumbrajs, O. Drumm, S. Illy, J. Jin, and M. Thumm. A 2-MW, 170-GHz coaxial cavity gyrotron. *Plasma Science, IEEE Transactions on*, 32(2):413–417, 2004.

[PEP+02] J. Petillo, K. Eppley, D. Panagos, P. Blanchard, E. Nelson, N. Dionne, J. DeFord, B. Held, L. Chernyakova, W. Krueger, S. Humphries, T. McClure, A. Mondelli, J. Burdette, M. Cattelino, R. True, K. T. Nguyen, and B. Levush. The michelle three-dimensional electron gun and collector modeling tool: theory and design. *Plasma Science, IEEE Transactions on*, 30(3):1238–1264, Jun 2002.

[Pet99] Michael I Petelin. One century of cyclotron radiation: Special issue on cyclotron resonance masers and gyrotrons. *IEEE transactions on plasma science*, 27(2):294–302, 1999.

[PHA+08] I. Gr. Pagonakis, J.-P. Hogge, S. Alberti, K. A. Avramides, and B. Piosczyk. Preliminary numerical study of the beam neutralization effect in the EU 170 GHz, 2 MW coaxial gyrotron. In *Plasma Science, 2008. ICOPS 2008. IEEE 35th International Conference on*, pages 1–1, June 2008.

[PHG+09] I. Gr. Pagonakis, J.-P. Hogge, T. Goodman, S. Alberti, B. Piosczyk, S. Illy, T. Rzesnicki, S. Kern, and C. Lievin. Gun design criteria for the refurbishment of the first prototype of the eu 170GHz/2MW/CW coaxial cavity gyrotron for ITER. In *Infrared, Millimeter, and Terahertz Waves, 2009. IRMMW-THz 2009. 34th International Conference on*, pages 1–2, Sept 2009.

[PIS+08] I. Gr. Pagonakis, S. Illy, M. Silva, J.P. Hogge, S. Alberti, K. A. Avramides, B. Piosczyk, F. Albajar, and T. Bonicelli. Parameterization technique for the preliminary gun design of the EU 170 GHz 1 MW conventional cavity gyrotron for iter. In *Infrared, Millimeter and Terahertz Waves, 2008. IRMMW-THz 2008. 33rd International Conference on*, pages 1–2, Sept 2008.

[PLY+12] Y. Pu, Y. Luo, R. Yan, G. Liu, and W. Jiang. Analysis of low frequency oscillations in magnetron injection guns. *Journal of Infrared, Millimeter, and Terahertz Waves*, 33(2):141–148, 2012.

[PNSA10] R. Pu, G. S. Nusinovich, O. V. Sinitsyn, and T. M. Antonsen. Effect of the thickness of electron beams on the gyrotron efficiency. *Physics of Plasmas*, 17(8), 2010.

[Poz09] D. M. Pozar. *Microwave engineering*. John Wiley & Sons, 2009.

[PPZ+16] I. Gr. Pagonakis, B. Piosczyk, J. Zhang, S. Illy, T. Rzesnicki, J.-P. Hogge, K. A. Avramidis, G. Gantenbein, M. Thumm, and J. Jelonnek. Electron trapping mechanisms in magnetron injection guns. *Physics of Plasmas*, 23(2), 2016.

[PTZ+13] E. Poli, G. Tardini, H. Zohm, E. Fable, D. Farina, L. Figini, N. B. Marushchenko, and L. Porte. Electron-cyclotron-current-drive efficiency in demo plasmas. *Nuclear Fusion*, 53(1):013011, 2013.

[PV04] I. Gr. Pagonakis and J. L. Vomvoridis. The self-consistent 3D trajectory electrostatic code ARIADNE for gyrotron beam tunnel simulation. In *Infrared and Millimeter Waves, 2004 and 12th International Conference on Terahertz Electronics,*

2004. Conference Digest of the 2004 Joint 29th International Conference on, pages 657–658, Sept 2004.

[RC70] V. L. Rideout and C. R. Crowell. Effects of image force and tunneling on current transport in metal-semiconductor (Schottky barrier) contacts. *Solid-State Electronics*, 13(7):993 – 1009, 1970.

[RC14] A. Roy Choudhury. *Investigations of after cavity interaction in gyrotrons including the effect of non-uniform magnetic field*. PhD thesis, Karlsruhe, 2014.

[RPK+10] T. Rzesnicki, B. Piosczyk, S. Kern, S. Illy, J. Jin, A. Samartsev, Andreas Schlaich, and M. Thumm. 2.2-MW record power of the 170-GHz european preprototype coaxial-cavity gyrotron for ITER. *Plasma Science, IEEE Transactions on*, 38(6):1141–1149, 2010.

[Rze15] T. Rzesnicki. personal communication, 2015.

[Sch15] A. Schlaich. *Time-dependent spectrum analysis of high power gyrotrons*. PhD thesis, Karlsruhe Institute of Technology, Karlsruhe, 2015.

[SIM72] R. Shimizu, T. Ikuta, and K. Murata. The monte carlo technique as applied to the fundamentals of epma and sem. *Journal of Applied Physics*, 43(10):4233–4249, 1972.

[Siv65] D. V. Sivukhin. *Reviews of plasma physics*. Consultants Bureau, New York, 1965.

[SJB+98] D. F. H. Start, J. Jacquinot, V. Bergeaud, V. P. Bhatnagar, G. A. Cottrell, S. Clement, L. G. Eriksson, A. Fasoli, A. Gondhalekar, C. Gormezano, G. Grosshoeg, K. Guenther, P. Harbour, L. D. Horton, A. Howman, H. Jackel, O. N.

Jarvis, K. D. Lawson, C. Lowry, M. Mantsinen, F. B. Marcus, R. Monk, E. Righi, F. G. Rimini, G. J. Sadler, G. R. Saibene, R. Sartori, B. Schunke, S. Sharapov, A. C. C. Sips, M. Stamp, and P. van Belle. D-T fusion with ion cyclotron resonance heating in the JET tokamak. *Physical Review Letters*, 80(21):4681–4684, 1998.

[SM71] R. Shimizu and K. Murata. Monte Carlo calculations of the electron-sample interactions in the scanning electron microscope. *Journal of Applied Physics*, 42(1):387, 1971.

[Spi13] Lyman Spitzer. *Physics of fully ionized gases.* Courier Corporation, 2013.

[TBPS83] J. C. Tonnerre, D. Brion, P. Palluel, and A. M. Shroff. Evaluation of the work function distribution of impregnated cathodes. *Applications of Surface Science*, 16(1):238–249, 1983.

[THP+01] T. J. J. Tala, J. A. Heikkinen, V. V. Parail, Yu. F. Baranov, and S. J. Karttunen. Itb formation in terms of flow shear and magnetic shears on jet. *Plasma Physics and Controlled Fusion*, 43(4):507, 2001.

[Thu03] M. Thumm. Progress in gyrotron development. *Fusion Engineering and Design*, 66:69–90, 2003.

[Thu11] M. Thumm. Progress on gyrotrons for ITER and future thermonuclear fusion reactors. *Plasma Science, IEEE Transactions on*, 39(4):971–979, April 2011.

[Thu15] M. Thumm. *State-of-the-art of high power gyro-devices and free electron masers.* KIT Scientific Reports 7693, Karlsruhe Institute of Technology, Karlsruhe, Germany, 2015.

[Tsi72] Sh. E. Tsimring. On the spread of velocities in helical electron beams. *Radiophysics and Quantum Electronics*, 15(8):952–961, 1972.

[Tsi01] Sh. E. Tsimring. Gyrotron electron beams: Velocity and energy spread and beam instabilities. *International Journal of Infrared and Millimeter Waves*, 22(10):1433–1468, 2001.

[VBV+02] T. J. Vink, A. R. Balkenende, R. G. F. A. Verbeek, H. A. M. van Hal, and S. T. de Zwart. Materials with a high secondary-electron yield for use in plasma displays. *Applied Physics Letters*, 80(12):2216–2218, 2002.

[VGGP12] M. V. Vilkov, M. Yu. Glyavin, A. L. Goldenberg, and M. I. Petelin. A magnetron injection gun with extraction of reflected electrons. *Technical Physics Letters*, 38(7):680–682, 2012.

[YAN10] R. Yan, T. M. Antonsen, and G. S. Nusinovich. Nonlinear analysis of low-frequency oscillations in gyrotrons. *IEEE Transactions on Plasma Science*, 38(6):1178–1184, June 2010.

[ZIPJ13] J. Zhang, S. Illy, I. Gr. Pagonakis, and J. Jelonnek. Preliminary study on the effects of emitter surface roughness on gyrotron electron beam quality. In *Infrared, Millimeter, and Terahertz Waves (IRMMW-THz), 2013 38th International Conference on*, pages 1–2. IEEE, 2013.

[ZT96] S. Zhang and M. Thumm. Kinetic description of the influence of electron beam misalignment on the performance of a coaxial cavity gyrotron. *Physics of Plasmas*, 3(7):2760–2765, 1996.

[ZZM06] N. A. Zavolsky, V. E. Zapevalov, and M. A. Moiseev. Influence of the energy and velocity spread in the electron beam on the starting conditions and efficiency of a gyrotron. *Radiophysics and Quantum Electronics*, 49(2):108–119, 2006.

Acknowledgements

The present work was accomplished during my doctoral studies at the Institute for Pulsed Power and Microwave Technology (IHM) at Karlsruhe Institute of Technology (KIT). Without the various support from China Scholarship Council (CSC), my home university NUDT, my supervisors, colleagues, friends and family it would not have been possible to perform this work.

At first, I do thank my dissertation supervisor Prof. Dr.-Ing. John Jelonnek for his confidence in my capabilities to perform this research work. He provided the chance to me for performing this research work under his guidance at the Institute for Pulsed Power and Microwave Technology (IHM). His professional guidance and encouragement have helped me to strengthen my skills as scientific researcher. What I have learned from him is more than only the hard skills to perform my work. He gives an example for the students of being trustful, high efficient and to be focused on a high quality on their work. I am sure, that I will benefit from that in my future professional career.

Besides my advisor, I would like to thank Prof. Dr. Alan Phelps, from the University of Strathclyde, UK, for being the second reviewer of my thesis. I am also grateful for the kind corrections and discussions he made to improve the thesis. Furthermore, my sincere thanks go to Prof. Dr. rer. nat. Dr. h.c. Manfred Thumm for his strong support, in particular his excellent suggestions and encouragement during this work. My work benefited significantly from his outstanding expertise in the field of vacuum electron devices. Already in my first week at IHM he provided me with his collections of publications on magnetron injection guns. These publications are the basis for this work.

I am deeply grateful to Dr. Stefan Illy for his continuous support and discussions for this work. He paid a lot of attention to my work. I do not remember how many times we have sat together to find solutions to the problems in this work. He is also very kind in helping me to overcome the personal difficulties I met during my research period at IHM.

Here, additional thanks go to Ioannis Pagonakis, Konstantinos Avramidis, Tomasz Rzesnicki and Martin Schmid for their significant help to this work, and the friendly and productive environment they provided to me. Special thanks go to the staff in the laboratory who always provided support for the experiments. At this place, I also thank Mrs. Martina Huber, Mrs. Wettstein and Mrs. Seitz of the administration office at IHM who were always very kind and helpful during my whole time at IHM.

In particular, my family and I are grateful to Dr. Jianbo Jin for his personal help during our stay at Germany.

Furthermore, I would like to thank my colleagues Joachim Franck, Sebastian Ruess, Chuanren Wu and Parth Kalaria for the great team spirit and memories we share together at IHM. Additional thanks goes to Andreas Schlaich for being a very helpful and kind roommate.

At China, there is a large number of people who helped me to start and accomplish this research work. In particular I do express my deepest gratitude to Prof. Huihuang Zhong, Prof. Jun Zhang and Prof. Hanwu Yang for their encouragement and support during the whole time of my Ph.D. work.

Particular thanks go to my family. I am deeply grateful for their continuous support. It was a tough time for my family to miss Dad. Here, very warm thanks go to my wife Hua Wang for her patience and taking care for our daughter Ziqi Zhang. Many thanks to you, Ziqi Zhang, you gave the spirit and power to me for performing my studies as fast and excellent as possible that I can be with you at China as soon as possible again.

Finally, I thank China Scholarship Council (CSC) for providing me the possibility to perform this excellent research study at KIT. Without their great support it would not have been possible to perform my doctoral studies at KIT.

Karlsruhe, August 2016 *Jianghua Zhang*

KARLSRUHER FORSCHUNGSBERICHTE AUS DEM
INSTITUT FÜR HOCHLEISTUNGSIMPULS- UND MIKROWELLENTECHNIK
(ISSN 2192-2764)

Die Bände sind unter www.ksp.kit.edu als PDF frei verfügbar oder als Druckausgabe bestellbar.

Band 1 **MATTHIAS BERINGER**
Design Studies towards a 4 MW 170 GHz Coaxial-Cavity Gyrotron. 2011
ISBN 978-3-86644-663-2

Band 2 **JENS FLAMM**
Diffraction and Scattering in Launchers of
Quasi-Optical Mode Converters for Gyrotrons. 2012
ISBN 978-3-86644-822-3

Band 3 **MATTIA DEL GIACCO**
Investigation of Fretting Wear of Cladding Materials in Liquid Lead. 2013
ISBN 978-3-86644-960-2

Band 4 **AMITAVO ROY CHOUDHURY**
Investigations of After Cavity Interaction in Gyrotrons
Including the Effect of Non-uniform Magnetic Field. 2013
ISBN 978-3-7315-0129-9

Band 5 **MICHAEL BETZ**
The CERN Resonant WISP Search (CROWS). 2014
ISBN 978-3-7315-0199-2

Band 6 **ANDREAS SCHLAICH**
Time-dependent spectrum analysis of high power gyrotrons. 2015
ISBN 978-3-7315-0375-0

Band 7 **DHIDIK PRASTIYANTO**
Temperature- and Time-Dependent Dielectric Measurements
and Modelling on Curing of Polymer Composites. 2016
ISBN 978-3-7315-0424-5

Band 8 **YIMING SUN**
Adaptive and Intelligent Temperature Control of Microwave
Heating Systems with Multiple Sources. 2016
ISBN 978-3-7315-0467-2

Band 9 **JIANGHUA ZHANG**
Influence of Emitter surface roughness and Emission inhomogeneity
on Efficiency and stability of high power Fusion gyrotrons. 2016
ISBN 978-3-7315-0578-5